CAMBRIDGE LIBRARY COLLECTION

Books of enduring scholarly value

Earth Sciences

In the nineteenth century, geology emerged as a distinct academic discipline. It pointed the way towards the theory of evolution, as scientists including Gideon Mantell, Adam Sedgwick, Charles Lyell and Roderick Murchison began to use the evidence of minerals, rock formations and fossils to demonstrate that the earth was older by millions of years than the conventional, Bible-based wisdom had supposed. They argued convincingly that the climate, flora and fauna of the distant past could be deduced from geological evidence. Volcanic activity, the formation of mountains, and the action of glaciers and rivers, tides and ocean currents also became better understood. This series includes landmark publications by pioneers of the modern earth sciences, who advanced the scientific understanding of our planet and the processes by which it is constantly re-shaped.

The Wonders of Geology

Gideon Mantell (1790–1852) was an English physician and geologist best known for pioneering the scientific study of dinosaurs. After an apprenticeship to a local surgeon in Sussex, Mantell became a member of the Royal College of Surgeons in 1811. He developed an interest in fossils, and in 1822 his discovery of fossil teeth which he later identified as belonging to an iguana-like creature he named *Iguanadon* spurred research into ancient reptiles. These volumes, first published in 1838, contain a series of eight lectures which describe and explain early principles of geology, stratification and fossil plants and animals to a non-scientific audience. These detailed volumes became Mantell's most popular work, and provide a fascinating view of the study of geology and palaeontology during the early nineteenth century. Volume 1 contains lectures 1–4, discussing the formation and composition of rock strata, chalk formations and fossil animals.

Cambridge University Press has long been a pioneer in the reissuing of out-of-print titles from its own backlist, producing digital reprints of books that are still sought after by scholars and students but could not be reprinted economically using traditional technology. The Cambridge Library Collection extends this activity to a wider range of books which are still of importance to researchers and professionals, either for the source material they contain, or as landmarks in the history of their academic discipline.

Drawing from the world-renowned collections in the Cambridge University Library, and guided by the advice of experts in each subject area, Cambridge University Press is using state-of-the-art scanning machines in its own Printing House to capture the content of each book selected for inclusion. The files are processed to give a consistently clear, crisp image, and the books finished to the high quality standard for which the Press is recognised around the world. The latest print-on-demand technology ensures that the books will remain available indefinitely, and that orders for single or multiple copies can quickly be supplied.

The Cambridge Library Collection will bring back to life books of enduring scholarly value (including out-of-copyright works originally issued by other publishers) across a wide range of disciplines in the humanities and social sciences and in science and technology.

The Wonders of Geology

Or, a Familiar Exposition of Geological Phenomena

VOLUME 1

GIDEON ALGERNON MANTELL
EDITED BY G.F. RICHARDSON

CAMBRIDGE
UNIVERSITY PRESS

CAMBRIDGE UNIVERSITY PRESS

Cambridge, New York, Melbourne, Madrid, Cape Town, Singapore,
São Paolo, Delhi, Dubai, Tokyo, Mexico City

Published in the United States of America by Cambridge University Press, New York

www.cambridge.org
Information on this title: www.cambridge.org/9781108021111

© in this compilation Cambridge University Press 2010

This edition first published 1838
This digitally printed version 2010

ISBN 978-1-108-02111-1 Paperback

THE COUNTRY OF THE IGUANODON RESTORED BY JOHN MARTIN ESQ KL

THE

WONDERS OF GEOLOGY;

BY

GIDEON MANTELL, LL.D. F.R.S.

AUTHOR OF

THE GEOLOGY OF THE SOUTH EAST OF ENGLAND,
ETC. ETC.

Silver Coins of Edward the First, in ironstone.—Page 58.

" To the natural philosopher there is no natural object unimportant
or trifling : from the least of nature's works he may learn the greatest
lesson."—SIR J. F. W. HERSCHEL.

" We know not a millionth part of the wonders of this beautiful
world."—LEIGH HUNT.

VOL. I.

LONDON:

RELFE AND FLETCHER, CORNHILL.

1838.

THE

WONDERS OF GEOLOGY;

OR,

A FAMILIAR EXPOSITION

OF

GEOLOGICAL PHENOMENA;

BEING THE SUBSTANCE OF

A COURSE OF LECTURES DELIVERED AT BRIGHTON.

BY

GIDEON MANTELL, LL.D. F.R.S.

FELLOW OF THE ROYAL COLLEGE OF SURGEONS;
AND OF THE LINNEAN AND GEOLOGICAL SOCIETIES OF LONDON AND CORNWALL;
HONORARY MEMBER OF THE PHILOMATHIC SOCIETY OF PARIS;
OF THE ACADEMIES OF NATURAL SCIENCES OF PHILADELPHIA; AND OF ARTS AND
SCIENCES OF CONNECTICUT; OF THE GEOLOGICAL SOCIETY OF PENNSYLVANIA;
OF THE PHILOSOPHICAL INSTITUTION OF BOSTON
OF THE HISTORICAL SOCIETY OF QUEBEC; AND OF THE PHILOSOPHICAL
SOCIETIES OF YORK, NEWCASTLE, ETC.

———◆———

FROM NOTES TAKEN BY G. F. RICHARDSON,
CURATOR OF THE MANTELLIAN MUSEUM, ETC.

VOL. I.

LONDON:
RELFE AND FLETCHER, CORNHILL.
1838.

TO

THE RIGHT HONOURABLE

GEORGE EARL OF MUNSTER,

F.R.S. F.G.S. &c. &c.

AS A TRIBUTE OF THE HIGHEST RESPECT AND REGARD,

𝔗𝔥𝔦𝔰 𝔙𝔬𝔩𝔲𝔪𝔢 𝔦𝔰 𝔍𝔫𝔰𝔠𝔯𝔦𝔟𝔢𝔡,

BY HIS LORDSHIP'S

MOST DEVOTED SERVANT,

GIDEON ALGERNON MANTELL.

CLAPHAM COMMON,
Feb. 27, 1838.

ADVERTISEMENT.

THIS Publication has originated in the wish
expressed by Dr. Mantell's auditors, to possess a
permanent record of his Lectures; while it has
at the same time been conceived, that a work
which should present a familiar exposition of the
principles and discoveries of Geology, encum-
bered with no larger share of technical details than
was absolutely required by the nature of the subject, would be acceptable to the public.

The Editor, in offering this explanation, feels
that he has discharged the duty imposed on him by

the slender tie which connects him with the work ; the pearls which it offers are not his own ; he has merely supplied the string which binds them together.

G. F. R.

Mantellian Museum,
Brighton.

TABLE OF CONTENTS.

VOL. I.

LECTURE I.

LECTURE II.

LECTURE III.

LECTURE IV.

TABLE OF CONTENTS.

VOL. II.

LECTURE V.

LECTURE VI.

LECTURE VII.

LECTURE VIII.

WONDERS OF GEOLOGY.

LECTURE I.

1. Introductory Remarks.—Having for many
years made fossil comparative anatomy my principal relaxation from the toils of a laborious and

B

extensive medical practice, the collection of organic remains which I had formed in the course of a considerable period, began to acquire an European celebrity through the writings of Cuvier, Brongniart, Humboldt, and other eminent savans, who had honoured my discoveries by their favourable notice. During the summer months, visitors to my collection became so numerous that I was compelled to limit the admission of strangers to certain days, when all were gratuitously admitted. This method was continued when I first took up my residence in Brighton; but a perseverance in this plan was soon found impracticable, from the impossibility of restricting visitors to the appointed hours: I was, therefore, obliged to close my collection to the public,—and hence the origin of the Sussex Royal Institution. As, prior to its establishment, I had, in compliance with the wishes of my friends, given several lectures on Geology, for charitable purposes, I was induced to undertake the delivery of discourses on the same subject, in the hope of promoting the interests of the infant Society. In conformity with this arrangement, I now enter upon the present course, which is designed to offer a familiar exposition of the philosophy of Geology. And permit me to observe, that my career as a lecturer will begin and end in Brighton; for at the termination of this session, should Providence allot me life and health, I shall remove to a less public, but not less important sphere of usefulness.

2. VALUE OF SCIENTIFIC PURSUITS.—It has been observed by a distinguished divine, that in order to obtain a proper sense of the importance of any science, and of the worth and beauty of the objects it embraces, nothing more is necessary than the intent and persevering study of them; and that such is the consummate perfection of all the works of the Creator, that every inquirer discovers a surpassing worth, and grace, and dignity, in that special department to which he has peculiarly devoted his attention. Whatever the walk of philosophy on which he may enter, that will be the path, which of all others will appear to him the most enriched, by all that is fitted to captivate the intellect, and excite the imagination. " Yet before we can attain that elevation from which we may look down upon and comprehend the mysteries of the natural world, our way must be steep and toilsome, and we must learn to read the records of creation in a strange language. But when this is once acquired it becomes a mighty instrument of thought; enabling us to link together the phenomena of past and future times, and giving the mind a domination over many parts of the natural world, by teaching it to comprehend the laws, by which the Creator has ordained that the actions of material things shall be governed."

3. IMPORTANCE OF GEOLOGY.—In the whole circle of the sciences, there is perhaps none that more strikingly illustrates the force and truth of these remarks, than geology; none whose language

is more mysterious, yet which offers to its votaries
rewards so rich, so wondrous, and inexhaustible. In
the shapeless pebble that we tread upon, in the rude
mass of rock or clay, the uninstructed eye would in
vain seek for novelty or beauty ; like the adven-
turer in Arabian story, the inquirer finds the cavern
closed to his entrance, and the rock refusing to give
up the treasures entombed within its stony sepulchre,
till the talisman is obtained that can dissolve the
enchantment, and unfold the wondrous secrets which
have so long lain hidden.

4. NATURE OF GEOLOGY.—Geology may be
termed the physical history of our globe,—it investi-
gates the structure of the planet on which we live,
and explains the characters and causes of the various
changes in the organic and inorganic kingdoms of
nature. It has been emphatically called, by the
most eminent philosopher of our time, the sister
science of Astronomy. But, relating as it does to
the history of the past, and carrying us back, by the
careful examination of the relics of former ages, to
periods so remote as to startle all our preconceived
opinions of the age of our globe, the fate of its early
cultivators has resembled that of the immortal
Galileo and the astronomers of his time ; and for a
similar reason, namely, the supposed discrepancy
between the discoveries and inferences of science,
and the Mosaic cosmogony.

5. HARMONY BETWEEN REVELATION AND GEO-
LOGY.—There was a time when every geologist was

called upon to defend himself against imputations of this kind ; but a more enlightened era has arrived, and it is unnecessary to allude to the circumstance, except to assure those who for the first time are called upon to follow the researches of the astronomer and the geologist, that in proportion as their minds become acquainted with the principles of scientific investigation, their apprehensions of any collision between the discoveries in the natural world, and the inspired record, will disappear. With regard to geology, I will content myself, on this occasion, with the following extract from the sermons of an eminent prelate, the Bishop of London : " As we are not called upon by Scripture to admit, so neither are we required to deny the supposition, that *the matter without form, and void, out of which this globe of earth was framed, may have consisted of the wrecks and relics of more ancient worlds,* created and destroyed by the same Almighty Power, which called our world into being, and will one day cause it to pass away."*

Thus, while the Bible reveals to us the moral history and destiny of our race, and teaches us that man and other living things have been placed but a few thousand years upon the earth, the physical monuments of our globe bear witness to the same truth; and as astronomy unfolds to us myriads of worlds, not spoken of in the sacred records, geology

* Sermons, by Dr. Charles James Blomfield, Bishop of London. 8vo. 1829.

in like manner proves, not by arguments drawn from analogy, but by the incontrovertible evidence of physical phenomena, that there were former conditions of our planet, separated from each other by vast intervals of time, during which this world was teeming with life, ere man and the animals which are his cotemporaries, had been called into being.

6. EXTENSIVE DURATION OF GEOLOGICAL PERIODS.—At the first step we take in geological inquiry, we are struck with the immense periods of time which the phenomena presented to our view must have required for their production, and the incessant changes which appear to have been going on in the natural world: but we must remember that time and change are great only with reference to the faculties of the being which notes them. The insect of an hour, contrasting its own ephemeral existence with the flowers on which it rests, would attribute an unchanging durability to the most evanescent of vegetable forms ; while the flowers, the trees, and the forests would ascribe an endless duration to the soil on which they grow : and thus, uninstructed man, comparing his own brief earthly existence with the solid framework of the world he inhabits, deems the hills and mountains around him coeval with the globe itself. But, with the enlargement and cultivation of his mental powers, he takes a more just, comprehensive, and enlightened view of the wonderful scheme of creation ; and while in his ignorance he imagined that the duration of the globe

was to be measured by his own brief span, and
arrogantly deemed himself alone the object of the
Almighty's care, and that all things were created
for his pleasure or his necessities; he now feels his
own dependence, entertains more correct ideas of
the mercy, wisdom, and goodness of the Creator;
and, while exercising his high privilege of being
alone capable of contemplating and understanding
the wonders of the natural world, he learns that
most important of all lessons,—to doubt the evidence
of his senses until confirmed by cautious and patient
investigation.

7. OBJECT OF THE PRESENT COURSE OF LEC-
TURES.—With these introductory remarks I proceed
to the consideration of the subjects selected for
the present lecture. And here I would observe,
that, from the magnitude and diversity of the objects
embraced by Geology, it is scarcely possible to offer,
in the brief space assigned to a course of popular
lectures, even an epitome of the wonders which
modern Geology has brought to light. Let this
consideration therefore be my apology for the hasty
or imperfect manner in which many interesting
phenomena may perhaps be noticed; and let me
also beg of you to consider that lectures of this
kind are intended to excite, rather than to satisfy, a
rational curiosity; that they are designed to pro-
mote a taste for philosophical pursuits, but cannot
supersede the necessity of study and of personal
investigation.

8. PHYSICAL GEOGRAPHY OF THE EARTH.—
The globe we inhabit may be described as a planetary
orb of a few thousand miles in circumference, and
of a spheroidal shape; its figure being such as a
body in a fluid state, and made to rotate on its axis,
would assume. Its mean density is five times greater
than that of water, its interior being double that of
the solid superficial crust: the internal part of the
earth, if cavernous, must therefore be composed of
very dense materials. Its surface is computed to
contain 190 millions of square miles, of which three-
fifths are covered by seas; another large proportion
by vast bodies of fresh water, and by polar ice and
eternal snows; so that taking into consideration
sterile tracts, morasses, &c., scarcely more than one-
fifth of the surface of the globe is fit for the habitation of man and terrestrial animals.* The area of
the Pacific Ocean alone is estimated as equal to the
entire surface of the dry land. The distribution of
the land is exceedingly irregular, the greater portion
being situated in the northern hemisphere, as a
reference to a map of the earth will clearly demonstrate. In a geological point of view, dry land
can only be considered as so much of the crust of
the earth as is above the level of the water, beneath
which it may again disappear. From accurate cal-
culations it is inferred that the present land might be
distributed over the bed of the ocean, in such manner

* Bakewell's Geology.

that the surface of the globe would present an uninterrupted sheet of water. We perceive, then, that
every imaginable distribution of land and water may
take place; and consequently, every variety of organic
life may find at different periods suitable abodes.

9. GEOGRAPHICAL DISTRIBUTION OF ANI-
MALS.—An investigation of the laws which govern
the distribution of animals and vegetables is an
inquiry of deep interest; but my limits compel me
to be brief, and as Mr. Lyell has treated the subject
in his accustomed lucid and masterly manner, I beg
to refer you to the third volume of his "Principles
of Geology," for more ample details. It will be
sufficient for my present purpose, to state that
although it might have been expected that, all other
circumstances being equal, the same animals and
plants would have been found in places of like
climate and temperature; yet this identity of distribution does not exist. When America was first
discovered, the indigenous quadrupeds were all dissimilar to those of the old world. The elephant,
rhinoceros, hippopotamus, giraffe, camel, horse,
buffalo, lion, tiger, &c. were not met with on the
new continent; while the American species of mammalia, as the llama, jaguar, paca, coati, sloth, &c.
were unknown in the old. New Holland contains,
as is well known, a most singular assemblage of
mammalia, consisting of more than forty species
of marsupial animals, of which the kangaroo is a
familiar example. The islands of the Pacific Ocean

contain no quadrupeds except hogs, dogs, rats, and
a few bats.

10. GEOGRAPHICAL DISTRIBUTION OF VEGE-
TABLES.—The distribution of vegetable life, although
perhaps more arbitrarily fixed by temperature and
by local influences than that of animals, presents
many anomalies. From numerous observations it
appears that vegetable creation took place in dif-
'ferent centres, each being the focus of a peculiar
species ; for many plants have a local existence, and
vegetate spontaneously in one district alone. The
cedar of Lebanon is indigenous to that mountain,
and does not grow spontaneously in any other part
of the world. But it will be sufficient in this place
to observe, that certain great divisions of the vegetable kingdom are distributed in certain regions ;
we shall have occasion to refer to this subject in
the lecture devoted to the consideration of Fossil
Botany.

11. TEMPERATURE OF THE EARTH.—The temperature of the globe is materially influenced by
solar light and heat; and hence the difference of
the seasons and of the climates of various latitudes. But there are also local causes, which
occasion great variations in its superficial temperature ; yet under equal circumstances the temperature
decreases from the tropics to the pole. There is
also an internal source of heat, the cause of which
has not yet been ascertained, but is probably
dependant on the original constitution of our planet.

12. NATURE OF THE CRUST OF THE EARTH.—
The greatest thickness of the superficial crust of the
globe, that is, of the mass of solid materials which
the ingenuity of man has been able to examine, is
estimated at about ten miles; this calculation extending from the highest mountain peaks to the
greatest natural or artificial depths. As the earth
is nearly eight thousand miles in diameter, the
entire series of strata hitherto explored, is but very
insignificant compared with the magnitude of the
globe; bearing about the same relative proportion
as the thickness of the paper which covers an artificial sphere, a foot in diameter. The inequalities
and crevices in the varnish of such an instrument,
would be equal in proportionate size to the highest
mountains and deepest valleys in the world. In this
diagram* (copied from Mr. De la Beche's admirable
work) the proportions are well displayed: thus a line
of an inch in breadth in the circle before you, represents a thickness of the external crust of the
earth equal to 100 miles; and these fine lines in the
margin, the altitude of the Alps, the Andes, and
the Himalayeh Mountains, the highest in the world.
As a thickness of 100 miles exceeds by ten times
that of the whole of the strata that are accessible
to human observation, we can understand how
disturbances of the earth's surface, even to ten

* To preserve as far as possible the language and spirit of
the original discourses, the references to diagrams and specimens are retained.—G. F. R.

times the depth of any of those which come within
the reach of our examination, may take place,
without in any sensible degree affecting the entire
mass of the globe. If these facts be duly considered,
the mind will be prepared to receive one of the most
startling propositions in modern geology—namely,
that the highest mountains have once been at the
bottom of the sea, and have been raised to their
present situations by subterranean agency,—some
gradually, others suddenly, but all, geologically
speaking, at a comparatively recent period.

13. COMPOSITION OF THE ROCKS AND STRATA.
—The materials which compose the superficial crust
of the earth, consist of numerous layers and masses
of earthy substances; of which combinations of
iron, lime, and silex (or the earth of flint), constitute a large proportion, the latter forming forty-five
per cent. of the whole. Those which have been
deposited the latest bear evident marks of mechanical origin, and are the water-worn ruins of older
rocks; as we descend, strata of a denser character
appear, which also exhibit proofs of having been
deposited by water; but when we arrive at the
lowest in the scale, a crystalline structure uniformly
prevails; and while in the former strata trees, plants,
shells, bones, and other remains of animals and
vegetables are found in profusion, in the lowermost
rocks all traces of organic life are absent.

14. CLASSIFICATION OF ROCKS.—In the infancy
of the science these remarkable phenomena gave

rise to an ingenious theory, which however, like all theories founded on insufficient data, has proved untenable. Still it may be convenient to notice the hypothesis, since the terms employed are still retained in the nomenclature of Geology. Agreeably to this theory, the mineral masses of which the crust of the earth is formed, are separated into three groups.

1st. *The Primitive* (now called *Primary*) *Rocks;* such as granite, sienite, porphyry, &c. : these are of crystalline structure, and evidently owe their present state to igneous agency. They are the lowermost rocks, and constitute the foundation, as it were, on which all the newer strata have been deposited; they also rise to the highest elevations on the surface of the globe. They were called primitive, because it was inferred, from the entire absence of organic remains, that they had been formed before the creation of animals and vegetables; but it is now clearly ascertained that granite and its associated rocks are, in fact, lavas of various ages.

2d. *The Transition Rocks.* These are superimposed on the primitive, and are more or less distinctly stratified—that is, are separable into layers, and contain the fossilized remains of animals, corals, plants, and shells. They were called transition, because it was assumed that they had been formed at the period when the surface of the earth and the seas were passing into a state fit for the reception of organized beings. Modern researches have however proved that they are strata altered by the effects of heat under high pressure.

3d. *The Secondary.* These have clearly originated from the destruction of the more ancient rocks, and have been deposited in hollows and depressions, by the action of rivers and seas. They abound in the mineralized remains of animals and plants; the most ancient enclosing zoophytes and shells; the next in antiquity containing, in addition, vegetable remains and fishes; those which succeed enveloping not only fishes, shells, zoophytes, and plants, but also bones of enormous reptiles. The chalk is the uppermost, or most recent of this class of strata. As the secondary rocks have manifestly been formed by the agency of water, it is clear that they were originally deposited in horizontal, or nearly horizontal layers or strata, although by far the greater portion have been broken up, and now lie in directions more or less inclined to the horizon.

For the convenience of study, this subdivision of the deposits is still retained, as will be hereafter shown. To the above groups modern geologists have added a fourth class.

4th. *The Tertiary.** These lie in hollows or basins of the chalk, and other secondary rocks, and are formed of the detritus of the ancient beds. They abound in shells, plants, zoophytes, crustacea,

* See Plate VI. Fig. 3.

fishes, &c.; and in them, with but one exception, the bones of mammalia first appear.

Of a later formation than the tertiary strata, are those accumulations of water-worn materials, which the surface of every country presents more or less abundantly. These are termed diluvial deposits; and in them are found the remains of existing species of animals, associated with those of others that are no longer to be met with on the face of the earth.

Even this slight examination of the strata affords convincing proofs of a former condition of animated nature, widely different from the present. We have evidence of a succession of periods of unknown duration, in which both the land and the sea teemed with forms of existence that have successively disappeared and given place to others; and these again to new races, approaching gradually more and more nearly to those which now inhabit the earth, till at length existing species make their appearance.

15. GEOLOGICAL MUTATIONS.—From this view of the physical structure of our planet we learn, at least so far as the limited powers of man can penetrate into the history of the past, that the distribution of land and water on its surface has been undergoing perpetual mutation; yet, that through countless ages the physical condition of the earth has not materially differed from the present; that the dry land has been clothed with vegetation, and tenanted by appropriate inhabitants; and that the sea and the bodies of fresh water have swarmed with living

forms ; that at a remote epoch, though animal and
vegetable life existed, the species were wholly different from any that now abound, and of a nature
fitted to live in a temperature much higher, and
more equally distributed, than could occur in the
present state of the earth ; and lastly, that in the inferior, or most ancient beds, all traces of mechanical
action, and of animal and vegetable organization, are
absent ; or in other words, have either never existed,
or *have been altogether obliterated.* Before entering upon the division of these discourses to which
the term Geology is commonly restricted, it will
facilitate our comprehension of many of the phenomena which the strata present to our notice, if in
this place we endeavour to penetrate the mystery
that veils the earliest condition of the earth ; and
which we shall in vain attempt, if our observations
are confined to our own planet.

16. CONNEXION OF GEOLOGY WITH ASTRONOMY.
— Here Geology conducts us to Astronomy, and
teaches us to look to the kindred spheres around us,
for the elucidation of the early history of our globe ;
and to consider our planet but as an attendant
satellite on a vast central luminary. The solar
system consists of the sun, whose mass is made up
of matter like our earth, surrounded by a luminous
atmosphere ; and of eleven small planets, which
revolve around it in various periods ; our earth being
the third in distance from the sun, and in bulk, as
compared with that body, of the size of a pea to

that of a globe two feet in diameter: and with a satellite, the moon, revolving round it. Upon examining the moon with powerful telescopes, we are enabled to ascertain that its surface is diversified by hill and valley; that it is a congeries of mountains, many of which are manifestly volcanic, some of the lava currents being distinctly visible. We have in fact a torn, crateriform, and disturbed surface, like that which we may conceive to have been presented by our earth, ere the pinnacles of the granite mountains were abraded, and the valleys neither smoothed nor filled up by the agency of water.* In Venus and Mercury the mountains appear to be enormous; while in Jupiter and Saturn there are but slight traces of any considerable elevations.

17. NEBULAR THEORY OF THE UNIVERSE.— Modern Astronomy instructs us that in the original condition of the solar system, the sun was the nucleus of a nebulosity or atmosphere, which revolved on its axis, and extended far beyond the orbits of all the planets ; the planets as yet having no existence. Its temperature gradually diminished, and becoming contracted by cooling, the rotation increased in rapidity, and zones of vapour or nebulosity were successively thrown off, the centrifugal force overpowering the central attraction ; the condensation of these separated masses constituting the planets and satellites. But this view of the conversion of gaseous matter

* See Appendix A.

c

into planetary bodies is not limited to our own system; it extends to the formation of the countless
myriads of suns and worlds which are distributed
throughout the universe! The sublime discoveries
of Sir Wm. Herschel have shown us that the realms
of space abound in nebulous bodies in every varied
condition, from that of a diffused nebulosity to suns
and worlds like our own. It must be admitted that
this assertion appears astounding,—and that it may
fairly be asked if man, the ephemeron of the material
world, can measure the mighty epochs which mark
the progressive development of suns and systems?
The genius of Herschel has effected this wonderful
achievement, and has explained the successive changes by which, through the agency of the eternal and
unerring laws of the Almighty, suns and worlds are
called into existence. By laborious and unremitting
observations, that distinguished astronomer, and his
no less gifted son, have demonstrated the progress of
nebular condensation,—not indeed from the phenomena presented by a single nebula, (for the process
can only become sensible through the lapse of hundreds, perhaps thousands, of years;) but by observations on the almost countless series of related,
contemporaneous objects in every varied state of
progression, from that of a cloud of luminous
vapour, to the most dense and mighty orbs that
appear in the firmament. As the naturalist in the
midst of a forest would be unable by a glance to
discover that the trees around him were in a state

of progressive change; yet, by perceiving that there were plants in different states of growth, from the acorn just bursting from the soil to the lofty oak that stood the monarch of the woods, could readily, from the succession of changes thus at once presented to his view, ascertain the progression of vegetable life, although extending over a period far beyond his own brief existence: in like manner, the astronomer looks into the wonders of the heavens, and by a survey of the sidereal world is able, by careful induction from the varied condition of the heavenly bodies around him, to discover the succession of changes, which, as regards a single nebula, even the duration of our solar system might possibly be insufficient to solve.   Thus it is that Herschel has traced, from nebular masses of absolute vagueness to others which present form and structure, the effects of the mysterious law which governs the stupendous stellular changes that are constantly taking place.

Some of these bodies appear as mere clouds of attenuated light—others as if curdling into separate masses—while many seem assuming a spheroidal figure.   Others again present a dense central nucleus of light surrounded by a luminous halo; and a series may thus be traced, from clusters of round bodies with increased central illumination, to separate nebulæ with single nuclei—to a central disk constituting a nebular star—and finally to an orb of light with a halo like our sun!

In the comets, those nebular bodies which belong
to our own and other systems, we have evidence
that even in the most diffused state of the luminous
matter, the masses which it forms are subservient
to the laws of orbicular motion : of which an interesting proof is afforded by Eucke's comet, that
mere wisp of vapour, which in a period but little
exceeding three years revolves around the central
luminary of our own system.   This beautiful theory
of Herschel and La Place is followed out by an
easy and evident process, through the formation
of planets and satellites, and explains the uniform
direction of their revolutions.   Yet not only is it
believed that such are the laws which the Creator
has established for the maintenance and government
of the universe, but it is satisfactorily shown, upon
mechanical principles, that such nebulæ *must of
necessity produce* planetary bodies.

18. FORMATION OF THE SOLAR SYSTEM.—In
our own system the formation of the planets and
satellites is thus explained.   The sun is a planetary
body with a nebulous atmosphere, the central nucleus
of a once extensive nebulosity.   During the condensation of this nebula the planets were successively
thrown off; the most distant, as Herschel, being the
first or most ancient, followed by Saturn, Jupiter,
the four asteroids, Mars, the Earth, Venus, and
Mercury; the satellites, as 'distinct worlds, being
the most recent of the whole.   In explaining their
formation, it is inferred that in any given state of

the rotating solar mass, the outer portion or ring might have its centrifugal force exactly balanced by gravity; but increased rotation would throw off that ring, which might sometimes retain its figure, of which we have a beautiful example in Saturn. This result, however, would not take place unless the annular band were of uniform composition, which would rarely be the case; hence the ring would most generally divide into several portions : these might sometimes be of nearly equal bulk, as in the asteroids; while in others they would coalesce into one mass. The solar nebulæ, thus thrown off at various periods, and constituting planets in a gaseous state, would each necessarily have a rotatory motion, and revolve in varying orbits around the central nucleus; and as refrigeration and consolidation proceeded, each might throw off entire annuli, or rings, or satellites, in like manner as the planets themselves had been projected from the sun.—But I must not pursue this most interesting subject farther; those who feel desirous of more ample information may consult a highly popular abstract of the discoveries of modern astronomy, recently published, under the title of "Views of the Architecture of the Heavens."*

You will at once perceive that this theory can in no wise affect the inference that the universe is the work of an all-wise and omnipotent Creator. "Let

* By Dr. Nichol, Professor of Practical Astronomy in the University of Glasgow.

it be assumed that the point to which this hypothesis guides us, is the ultimate boundary of physical science—that the nearest glimpse we can attain of the material universe, displays it to us occupied by a boundless abyss of brilliant matter; still we are left to inquire how space became thus occupied—whence matter thus luminous? And if we are able to establish by physical proofs, that the first fact which the human mind can trace in the history of the heavens is, that 'there was light,' we are irresistibly led to the conclusion, that ere this could take place 'GOD SAID, *Let there be light.*' "

This theory of the condensation of nebular matter into suns and worlds, marvellous as it may appear, will be found on due reflection to offer the only rational explanation of the phenomena observable in the sidereal heavens, and in our own globe; and its beautiful simplicity is in correspondence with the unity of design so manifest throughout the works of the Eternal.

19. GASEOUS STATE OF THE EARTH.—Though the mind unaccustomed to philosophical inquiries, may find it difficult to comprehend the idea that this planet once existed in a gaseous state, this difficulty will vanish upon considering the nature of the changes that all the materials of which it is composed must constantly undergo. Water offers a familiar example of a substance existing on the surface of the globe, in the separate states of rock, fluid, and vapour; for water consolidated into ice

is as much a rock as granite or the adamant, and, as we shall hereafter have occasion to remark, has the power of preserving for ages the animals and vegetables that may be therein imbedded. Yet, upon an increase of temperature, the glaciers of the Alps, and the icy pinnacles of the Arctic circles, disappear; and, by a degree of heat still higher, might be resolved into vapour; and by other agencies might be separated into two invisible gases—oxygen and hydrogen. Metals may in like manner be converted into gases; and in the laboratory of the chemist, all kinds of matter easily pass through every grade of transmutation, from the most dense and compact to an aeriform state. We cannot, therefore, refuse our assent to the conclusion, that the entire mass of our globe might be resolved into a permanently gaseous form, merely by the dissolution of the existing combinations of matter.

From the light thus shed by modern Astronomy upon many of the dark and mysterious pages of the earth's physical history, we learn that the dynamical changes which have taken place in our globe—all the wonderful transmutations of its surface revealed to us by geological investigations—may be referable to the operation of the one, simple, universal law, by which the condensation of nebular masses into worlds, through periods of time so immense as to be beyond the power of human comprehension, is governed.

The internal heat of our globe—the evidence afforded by fossil organic remains of a more equally diffused and higher temperature of the surface in the earlier state of the earth—and the elevatory process that has taken place and is still in active operation—all refer to such an origin, and such a constitution of our planet, as that contemplated by the nebular theory. This elevatory process is not peculiar to our own planet; for, as we have elsewhere remarked, Venus, Mercury, the Moon, and perhaps the Sun itself, exhibit evidence of a similar where remarked, Venus, Mercury, the Moon, and perhaps the Sun itself, exhibit evidence of a similar action.* In a philosophical point of view, the present physical epoch of the earth " is that of the nical and chemical action upon the previously consolidated materials."†

20. METEORITES.—Intimately connected with this division of our subject, is the remarkable phenomenon of the fall of foreign bodies, called meteorites or meteoric stones, on our earth. The specimen before me, for which I am indebted to my kind and distinguished friend, Professor Silliman, of Yale College, Connecticut, is the fragment of a mass which was seen to fall at Nanjenoy, in Maryland, North America, a few years since. The

* This subject is treated at large in the interesting work of M. De la Beche, entitled " Researches in Theoretical Geology."
† Dr. Nichol.

following description by an eye-witness of the descent
of this body, will serve to illustrate the ordinary
phenomena which attend the appearance of these
mysterious visitors.*

"On the 10th of February, between the hours of
twelve and one o'clock, I heard an explosion, as I
supposed of a cannon, but somewhat sharper. I immediately advanced with a quick step about twenty
paces, when my attention was arrested by a buzzing
noise, as if something was rushing over my head,
and in a few seconds I heard something fall. The
time which elapsed from my first hearing the explosion to the falling, might have been fifteen seconds.
I then went with some of my servants to find where
it had fallen, but did not at first succeed; however,
in a short time the place was found by my cook,
who dug down to it; and a stone was discovered
about two feet below the surface. It was sensibly
warm, and had a sulphurous smell: was of an
oblong shape, and weighed sixteen pounds and
seven ounces. It has a hard vitreous surface. I
have conversed with many persons, living over an
extent of perhaps fifty miles square: some heard the
explosion, while others heard only the subsequent
whizzing noise in the air; all agree in stating that
the noise appeared directly over their heads. The
day was perfectly fine and clear. There was but
one report heard, and but one stone fell, to my

* American Journal of Science.

knowledge; there was no peculiar smell in the air:
it fell within 250 yards of my house."*

That ornament and pride of her sex, Mrs. Somerville, has the following interesting remarks on this
subject:—"So numerous are the objects which meet
our view in the heavens, that we cannot imagine a
part of space where some light would not strike the
eye: innumerable stars—thousands of double and
multiple systems—clusters in one blaze with their
ten thousands of stars—and the nebulæ amazing us
by the strangeness of their forms; till at last, from
the imperfection of our senses, even these thin and
airy phantoms vanish in the distance. If such
remote bodies shone by reflected light, we should
be unconscious of their existence; each star must
then be a sun, and may be presumed to have its
system of planets, satellites, and comets, like our
own; and for aught we know, myriads of bodies
may be wandering in space, unseen by us, of whose
nature we can form no idea, and still less of the
part they perform in the economy of the universe.
Nor is this an unwarranted presumption: many
such do come within the sphere of the earth's attraction, are ignited by the velocity with which they

---

* An analysis of this aerolite gave the following results:—

Oxide of Iron . . . . 24.
Nickel . . . 1.25
Silica with earthy matter 3.46
Sulphur, a trace.
———
28.71

pass through the atmosphere, and are precipitated with great violence to the earth. The fall of meteoric stones is much more frequent than is generally believed: hardly a year passes without some instances occurring; and if it be considered that only a small part of the earth is inhabited, it may be presumed that numbers fall into the ocean, or on the uninhabited parts of the land, unseen by man. They are sometimes of great magnitude: the volume of several has exceeded that of the planet Ceres, which is about seventy miles in diameter. One which passed within twenty-five miles of us was estimated to weigh about *six hundred thousand* tons, and to move with a velocity of about twenty miles in a second—a fragment of it alone reached the earth. The obliquity of the descent of meteorites, the peculiar substances of which they are composed, and the explosion attending their fall, show that they are foreign to our planet. Luminous spots, altogether independent of the phases, have been seen on the dark parts of the moon; these appear to be the light arising from the eruption of volcanoes; whence it has been supposed that meteorites have been projected from the moon by the impetus of volcanic eruption. If a stone were projected from the moon in a vertical line with an initial velocity of 10,992 feet in a second—a velocity but four times that of a ball when first discharged from a cannon—instead of falling back to the moon by the attraction of gravity, it would come within the sphere of the earth's attraction, and revolve around it like a satellite. These bodies, impelled either by the direction of the primitive impulse, or by the disturbing action of the sun, might ultimately penetrate the earth's atmosphere and arrive at its surface. But from whatever source meteoric stones may come, it is highly probable that they have a common origin, from the uniformity, we may almost say identity, of their chemical composition."*

Von Hoff, in an admirable essay on the origin of meteoric stones, † observes, that although it is demonstrated mathematically, that aerolites and masses of native iron which fall from the air, *may* be derived from the moon, yet the weight of evidence is in favour of their being nebulous matter suddenly condensed, and which descends to this planet's surface when this mysterious process takes place within the sphere of the earth's attraction. These masses present a general correspondence in their structure and appearance, having (with the exception of native iron) a crystalline character internally, and a black slaggy crust externally, as is seen in this specimen from Nanjenoy.

Assuming then that our planet, when first called into being by the fiat of the Creator, was a gaseous mass "without form and void," and destined through countless ages to undergo mutations which were

* Connexion of the Physical Sciences, p. 423. 4th Edition.
† A Translation of this Memoir appeared in Jameson's Edinburgh New Philosophical Journal, July 1837.

designed ultimately to prepare it for the abode of the human race, we proceed to investigate the causes and effects of those agencies by which its surface is still modified. The consideration of what Sir John Herschel so emphatically terms "that mystery of mysteries"—the first appearance of organic life on our globe, will be reserved for the concluding lecture.

21. EXISTING GEOLOGICAL CHANGES.—In this division of the subject, it will be my object to explain in a clear and familiar manner some of those physical changes which, unheeded or unappreciated, are taking place around us; and which, operating on a large scale, and through a long period of time, are capable of producing effects that materially modify the earth's surface, and give rise to phenomena which, when viewed in the aggregate, fill the uninformed mind with astonishment, and cause it to call up imaginary convulsions and catastrophes to explain the result of some of the most simple operations of nature. As the mere lines that compose the alphabet constitute, when placed in combination, the mighty engine by which the master spirits of our race enlighten and benefit mankind; so natural causes, in themselves apparently inadequate to produce any important effects, become, by their combined and continued operation, an irresistible power, converting the dry land into the bed of the ocean, and the bed of the ocean into dry land; thus fulfilling that universal law of the

Creator, which subjects every particle of matter to alternate decay and renovation.

Before proceeding farther in this inquiry, I would notice an opinion, so generally prevalent that it may possibly be entertained by some present; namely, that the phenomena which will come under our consideration, have been produced by that miraculous event, the deluge recorded in Scripture. Now whatever may have been the modifications of the earth's surface produced by that catastrophe, they must on the present occasion be wholly excluded from our consideration; for, as we shall hereafter perceive, the changes to which geological inquiries relate are of a totally different character.

I have now to direct your attention to those natural operations which, when properly investigated, will afford an easy explanation of facts of the highest interest and importance; will teach us how this limestone has been formed of brittle shells, and this marble filled with the coral to which it owes its beautiful markings—how wood has been changed into stone; and plants and fishes have become enclosed in the solid rock. I wish to explain to you that the ground on which we stand was not always dry land, but once formed the bottom of a sea or an estuary—that the hills, now so smooth and rounded, and clothed with beautiful verdure, have been formed in the profound depths of the ocean, and may be regarded as vast tumuli, in which the remains of beings that lived and died in the early ages of the globe are entombed;—and that the weald of Kent and Sussex, that rich and cultivated district which fills up the area between the chalk hills of Sussex, Surrey, Kent, and Hampshire, was once the delta of a mighty river, that flowed through a country which is now swept from the face of the earth—a country more marvellous than any that even romance or poetry has ventured to pourtray.

22. Effects of Streams and Rivers.—In pursuance of this object I shall first take into consideration the action of running water—of streams, and rivers. I need not dwell on those meteorological causes by which the descent of moisture on the surface of the earth is regulated; but shall content myself with observing, that rivers are the great natural outlets which convey the superfluous moisture of the land, into the grand reservoir, the ocean. And so exactly is the balance of expenditure and supply maintained, that all the rivers on the face of the earth, though constantly pouring their mighty floods into the ocean, do not affect its level in the slightest perceptible degree. Hence we may assume that the quantity of moisture evaporated from the surface of the sea, is exactly equal to the sum of all the water, in all the rivers in the world. But although the body of fresh-water poured by the rivers into the basin of the sea is again displaced by evaporation, yet there is an operation silently and constantly going on, which becomes an agent of

universal change. The rivulets issuing from the
mountains are more or less charged with earthy
particles, worn from the rocks and strata over which
they flow: their united streams in their progress
towards the rivers become more and more loaded
with adventitious matter; and as the power of
abrasion becomes greater, by the increase in the
quantity and density of the mass of water, a large
proportion of materials is mechanically or chemically suspended in the fluid, and carried into the
bed of the ocean. If the current be feeble, much
of the mud, and the larger pebbles, will be thrown
down in the bed of the river—hence the formation
of the alluvial plains in the valleys of the Arun, the
Adur, the Ouse, and Cuckmere, in this county.
But the greater portion will be transported to the
mouths of the rivers, and there form those accumulations of the fluviatile spoils of the land which constitute deltas and estuaries; the finest particles,
however, will be carried far into the sea, and,
transported by currents and agitated by the waves,
will at length be precipitated into the profound and
tranquil depths of the ocean.   But the waters
convey not only the mud and water-worn materials
of the country over which they flow: leaves,
branches of trees, and other vegetable matter—and
the remains of the animals that fall into the streams,
with shells and other exuviæ—human remains, and
works of art—are also constantly transported and
imbedded in the mud, and silt, and sand of the

delta ; some of these remains being occasionally
drifted out to sea, and deposited in its bed.

23. DELTAS OF THE GANGES, AND MISSISSIPPI.
—The changes here contemplated, as they are going
on in our own island, may appear insignificant, and
incapable of producing any material effect on the
earth's surface; but if we trace the results in
countries where these agents operate on a larger
scale, we shall at once perceive their importance,
and that time only is wanting, to form accumulations of strata, equal in extent, and of precisely
the same characters with many of those ancient
deposites, which will hereafter come under our observation.

Mr. Lyell states, that from experiments made with
great care, it has been ascertained that the quantity
of solid matter brought down by the Ganges
and carried into the sea annually, is equal to
6,368,077,440 tons : in other words, to a mass of solid
materials, equal in weight to sixty of the great pyramids of Egypt; the base of the great pyramid being
eleven acres, and the perpendicular height 500 feet.
The Burrampooter, another river in India, conveys
annually as much earthy matter into the sea as the
Ganges. The waters of the Indus, as the celebrated
traveller, Captain Burns, informed me, are alike
loaded with earthy materials.

In the mighty rivers of America, the same effects
are observable ; the immense quantities of trees
brought down by the Mississippi and imbedded in

D

its deposites are almost incredible, and the basin of
the sea around the embouchure of that river, is becoming shallower every day, by the sole agency of
the operation now under our consideration. In the
sediments of these rivers, the animals as well as the
plants of the respective countries are continually
enveloped. It is therefore evident, that should these
deltas become dry land, the naturalist could, on
examination of the animal and vegetable remains
imbedded in the fluviatile sediments, readily determine the characters of the fauna, and flora of the
countries through which the rivers had flowed. We
may here observe, that in tropical regions, where
animal life is profusely developed, and but little
under the control of man, the animal remains buried
in deltas, are far more abundant than in those of
European countries, which are thickly peopled, and
in a high state of civilization. The enterprising, but
unfortunate Lander informed me, just before he
embarked on his last fatal expedition to Africa, that
in many parts of the Quorra, or Niger, the bed of
that river, so far as the eye could reach, teemed with
crocodiles and hippopotami ; and that so great was
their number, that he was oftentimes obliged to drag
his boat on shore lest it should be swamped by
these animals. But it is unnecessary for me to dwell
longer on these operations, which are so admirably
elucidated in the work of my friend Mr. Lyell :
it will suffice to have shown, that by the simple
operation of running water, great destruction and

modification of the surface of the land are everywhere taking place ; and that at the same time, accumulations of fluviatile deposits are forming on an
extensive scale, and enveloping animal and vegetable
remains. Thus, in the deltas of the rivers of this
country, we find the bones and antlers of the deer,
horse, and other domesticated animals, with the
trunks and branches of trees and plants, of our island,
and river and land shells, and bones of man, and
fragments of pottery, and other works of art : while
in those of the Ganges, and the Nile, the remains
of the animals and vegetables of India, and of Egypt,
are respectively entombed.

There is one circumstance connected with these
phenomena which it will be necessary here to
consider. You well know that the quantity of
water in streams and rivers, varies considerably
at different periods of the year ; that in the rainy
season the bed of the river is overflowing, and the
waters remarkably turbid: the depositions, therefore,
must be much greater at those periods than in the
summer months, when the streams are feeble, and the
rivers shallow. We must also remember, that in
that part of the rivers affected by the tides, there is
a constant flux and reflux of the waters, and from
these causes the depositions must, in a certain
degree, be periodical. Accordingly we find them
disposed in strata or layers, from the partial consolidation of the surface of one layer of mud, before
the superincumbent layer was precipitated upon

it.* Where a river terminates in an extensive estuary, the sea throws over the layer of mud brought
down by the river, a covering of sand : and frequently
these alternate with the greatest regularity, the receding of the tide allowing the fresh water to deposit its
mud, and the advance of the sea discharging sand
over the surface.

24. RIPPLED SAND.—And here we may notice
another phenomenon. Every one must have observed,
when walking by the banks of a river at low water,
or on the sands of the sea-shore, that when the water
has been agitated by the wind, the surface of the
mud, or sand, is undulated, or furrowed over by
the rippling of the waves ; the ripple marks presenting various appearances, according to the force
and direction of the currents. Frequently too, the
vermes, and molluscous animals, mark the surface
with meandering lines, and ridges ; and these varied
markings on the sand are preserved, if a thin pellicle
of mud be deposited over them before the next advance

* An American gentleman, who visited Egypt in 1834, remarks, in a letter to my distinguished friend Professor Silliman,
that wherever a fresh break takes place in a bank of consolidated mud, in the delta of the Nile, it is easy to trace the deposites
of each successive year, by means of the lighter earth on the top
of each. When a portion is taken into the hand, it separates
at those lines into layers ; and on closely examining the edges of
these, very delicate thin lines are perceptible, showing a laminated structure, like those observable in the coal-shales. Judging
from these layers, the annual deposites appear to vary considerably, but the average thickness is little more than a quarter
of an inch.

of the waves. I shall have occasion to refer to these appearances hereafter. We must also remark that there are certain kinds of mollusca, or shell-fish, that can only live in fresh water; others that are confined to the sea; while a third class is restricted to the brackish waters of estuaries. Accordingly, in the deposites under our consideration, the river and estuary species are abundant, while the marine only occur as stragglers, and are comparatively rare. Land plants, and those which affect a marshy soil, as the equiseta or mare's-tails, reeds, and rushes, are likewise often accumulated in such quantities as to form beds of peat.

25. LEWES LEVELS.—It will serve to impress the subject more forcibly upon our minds if we refer to some local example of fluviatile deposites: and from its immediate vicinity to Brighton, I select the valley of the Ouse, between Newhaven and Lewes, which is one of several estuaries from whence the sea has retired within the last eight or ten centuries. This valley is bounded by an amphitheatre of chalk hills; the river enters it through a gorge of the Downs on the north, and, pursuing a tortuous course through the valley, discharges its waters at Newhaven. This alluvial plain is called Lewes Levels; and here and there is flanked by headlands, and ancient cliffs, while a few insular mounds of chalk rise up through the fluviatile depositions, which have been accumulating during a long period of time. The following diagram represents

a section of the valley of the Ouse, from east to west.

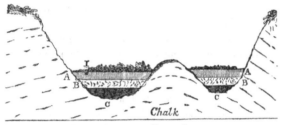

TAB. 1. SECTION OF LEWES LEVELS.

Here we have a depression (or basin, as it is termed by geologists) of the chalk, partially filled up by layers of indurated mud or silt; the surface being clothed with verdure, and the bed of the river (I). situated near the eastern chalk cliffs. The deposits which repose on the chalk are as follows:—

A bed of peat about five feet in thickness: formed of decayed twigs and leaves of the hazel, oak, birch, &c. enclosing trunks of large trees.

A.A.—Blue clay, or indurated mud, containing several species of fresh-water shells, like those which now inhabit the river or ditches; with numerous *indusiæ*, or cases of the larvæ of *Phryganeæ*, or caddis-worms. Bones of the horse, and deer, also occur in the lower part of this bed.

B.B.—Clay, containing fresh-water shells, with an intermixture of existing marine species, as the common cockle, (*Cardium edule*,) tellina, &c.

C. C.—Blue clay, inclosing marine shells, viz. cockles, muscles, &c. without any intermixture of fluviatile species. In this bed a skull of the narwal, or sea-unicorn (*Monodon monoceros*,) and porpoise have been discovered.

From the nature of these deposites we learn that this valley was once an arm of the sea, and that the following is the sequence of the physical changes which have taken place :—

First, There was a salt-water estuary, inhabited by marine shell-fish of the same species as those now existing in the British Channel ; and into this estuary Cetacea occasionally entered.

Secondly, The inlet grew shallower, the water brackish, and marine and fresh-water shells were mingled in its blue argillaceous sediment.

Thirdly, The shoaling continued until fresh-water so much predominated, that fluviatile shells, and aquatic insects, could alone exist.

Fourthly, A peaty swamp, or morass was formed, by the drifting of trees, and plants, from the forest of Andreadswald, which formerly occupied the weald of Sussex ; and terrestrial quadrupeds were occasionally imbedded.

Lastly, The soil, being inundated by the land floods only at distant intervals, became a verdant marshy plain.*

* See Mr. Lyell's Principles of Geology, fifth edition, vol. III. p. 265. Geology of the South-east of England, p. 16.

26. REMAINS OF MAN IN MODERN ALLU-
VIUM.—But the fluviatile deposites in the river
valleys of the South Downs often contain not only
the bones of the deer, horse, boar, and other terrestrial animals, but also human skeletons, sometimes
in coffins of exceedingly rude workmanship; and
canoes,* and other remains of the early inhabitants

---

* *Ancient British Canoe.* In 1835 a canoe was discovered at
the depth of several feet in a bed of silt, occupying an ancient
branch of the river Arun, at North Stoke, near Arundel. It
has been presented, by my noble friend the Earl of Egremont,
to the British Museum; and is placed on the right hand of the
entrance of the court.   This canoe is nearly thirty-five feet in
length, four and a half wide in the centre, three feet three inches
broad at one extremity, and two feet ten inches at the other; and is
about two feet deep. It is formed of the single trunk of an oak,
which has been hollowed out and brought to its present shape
with great labour; it is evidently the workmanship of a very
early period, and in all probability was constructed by some of
the earliest inhabitants of our island, before the use of iron or
even brass was known: the original tree must have been fifteen
or sixteen feet in circumference.   Three projections, left in the
interior of the boat, appear to have been designed for seats; it
is manifest therefore that the persons who constructed this vessel
were unacquainted with the art of forming boards.   This canoe
is so similar to some of those which were fabricated by the
aborigines of North America, when first visited by Europeans,
that we can have no hesitation in concluding that it was constructed in a similar manner; namely, by charring such portions
of the tree as were to be removed, and scooping them out with
stone instruments: no doubt this canoe belongs to the same
period as the flint and stone instruments called *celts*, which are
found in the tumuli on the South Downs; it is now in the state
of peat or bog-wood.

of our island. This human skull, for which I am
indebted to my intelligent friend, Warren Lee, Esq.
of Lewes, was dug up at a great depth in the blue
silt of Beeding Levels; it was inclosed, together with
the other bones of the skeleton, in a coffin of oak,
which was evidently of high antiquity, being formed
of four planks, or rather squared trunks of trees,
held together by oaken pegs. This skull is of
a dark bluish-brown colour, like the bones of the
deer and horse of similar deposites. This appearance
is owing to an impregnation of iron; when first dug
up, blue phosphate of iron filled up the interstices of
the bones. The state of the teeth is remarkable;
they are worn down almost smooth, although the
individual must have been in the prime of life; a
fact which seems to indicate that grain, or some
other hard substances, constituted a large proportion
of his customary food.

27. PEAT BOGS.—Before proceeding to the next
subject, I will advert to those extensive accumulations of vegetable matter called Peat Bogs; which
are morasses, covered with successive layers or beds
of mosses, reeds, equiseta, rushes, and other plants
that affect a marshy soil; and in particular of a kind
of moss, the *Sphagnum palustre*, which generally
constitutes a large proportion of the entire mass.
The beds of peat are annually augmented by the
peculiar mode of increase of the peat-moss, which
throws up a succession of shoots to the surface,
while the parent plants decay and form a new

layer of the soil; a process analogous, as we shall explain hereafter, to the mode of increase of the coral reefs.

The peat bogs of Ireland are of great extent: one of the mosses on the banks of the Shannon, in breadth two or three miles, is fifty miles in length. Mr. Lyell (to whose admirable work I must again refer you for more ample information on the effects of modern causes) observes, that the peat mosses of the North of Europe occupy the scite of the ancient forests of oak, and pine; and that the fall of trees from the effect of storms, or natural decay, by obstructing the draining of a district, and thus giving rise to a marsh, occasions the origin of most of the peat bogs. Mosses, and other marsh plants spring up, and soon overwhelm, and bury the prostrate forests. Hence the occurrence of trunks and branches of enormous oaks, firs, &c. with their fruits.

De Luc states, that the scite of many of the aboriginal forests on the Continent, is now covered by mosses and fens, and that many of these changes are attributable to the destruction of the forests by the Romans. A remarkable fact relating to peat bogs must not be omitted; namely, the occasional occurrence of the bodies of men and animals, in a high state of preservation, at a great depth. In some instances the bodies are converted into a fatty substance, resembling spermaceti, called *adipocire*.

28. SUBTERRANEAN FORESTS.—Independently

of the trees immersed in peat bogs and morasses, there are also found entire forests buried deeply in the soil; the trees having their roots, trunks, branches, fruits, and even leaves, more or less perfectly preserved. Several accumulations of this kind have been discovered on the coast of Sussex, occupying low alluvial tracts, that are still subject to periodical inundations. * The trees are chiefly of the oak, hazel, fir, birch, yew, willow, and ash; in short, almost every kind that is indigenous to this island occasionally occurs. The trunks, branches, &c. are dyed throughout of a deep ebony colour by iron; the wood is firm and heavy, and sometimes sufficiently sound for domestic use. In Yorkshire it is employed in the construction of houses. The specimens which I now place before you, (and for which I am indebted to my distinguished friend Professor Babbage) exhibit the usual character of such remains; they are portions of the trunks of large trees of the yew, oak, and fir.

29. GEOLOGICAL EFFECTS OF THE SEA.— While the mountains, valleys, and plains of the interior of a country, are undergoing slow but perpetual change by the combined effects of atmospheric agency, and of running water, the coasts, and shores, are exposed to destruction from the action of the waves, and the encroachments of the sea. When the land presents a high and rocky coast, the waves

* See the Fossils of the South Downs.

by their incessant action undermine the cliffs, which
at length fall down, and cover the shore with their
ruins. The softer parts of the strata, as the chalk,
marl, clay, &c. are rapidly disintegrated and washed
away : while the more solid materials are broken,
and rounded, by the continual agitation of the water,
and form those accumulations of beach and sand
which line our shores, and serve, in some situations,
to protect the land from further encroachments.
But when the cliffs are entirely composed of soft
substances, their destruction is very rapid, unless
artificial means are employed for their protection ;
and these in many instances are wholly ineffectual.

The encroachments of the ocean upon the land
effected by this operation, often give rise to sudden
and extensive inundations, and the destruction of
whole tracts of country. Along the Sussex coast the
inroads of the sea have been noticed in the earliest
historical records ; and you are doubtless aware, that
the scite of the ancient town of Brighton, is entirely
swept away, the sands, and the waves, now occupying the tract where the first settlers on this coast
fixed their habitation.* On low and sandy coasts,
the waves drive the loose and lighter materials
towards the shore; and the drifted sand, becoming
dry at the reflux of the tide, is carried by the wind
inland ; and in some situations is accumulated in
such quantities as to form ranges of hills, which in

* See Geology of the South-East of England, p. 23.

their progress overwhelm fertile tracts, and engulf churches, and even entire villages. These sandbanks or downs, loose and fluctuating as they are in their first stage of advancement, become under certain circumstances, fixed, and converted into solid stone—a process to which we shall presently advert.

30. BED OF THE OCEAN.—But the production of beach, and gravel, and sand, on the shores, and the drifting of sand inland, are effects far less important than those which are going on in the profound depths of the ocean. In the tranquil bed of the sea, the finer materials, held in mechanical or chemical suspension by the waters, are precipitated and deposited, enveloping and imbedding the inhabitants of its waters, together with the remains of such animals, and végetables of the land, as may be floated down by the streams and rivers. But, in the beautiful language of Mrs. Hemans,—

"The depths have more! What wealth untold
  Far down and shining through their stillness lies!
They have the starry gems, the burning gold,
  Won from a thousand royal argosies!

"Yet more—the depths have more! Their waves have roll'd
  Above the cities of a world gone by—
Sand hath filled up the palaces of old,
  Sea-weed o'ergrown the halls of revelry.

"To them the love of woman hath gone down,—
  Dark flow their tides o'er manhood's noble head,
O'er youth's bright locks, and beauty's flowery crown."—

Yes! in these modern depositions, the remains of man, and of his works, must of necessity be continually engulfed, together with those of the animals which are his contemporaries.

Of the nature of the bed of the ocean, we can of course know but little from actual observation. Soundings, however, have thrown light upon the deposites now forming in those depths, which are accessible to this mode of investigation, and thus we learn, that in many parts immense accumulations of the wreck of testaceous animals, intermixed with sand, gravel, and mud, are going on. Donati ascertained the existence of a compact bed of shells, 100 feet in thickness, at the bottom of the Adriatic, which in some parts was converted into marble. In the British Channel, extensive deposites of sand, imbedding the remains of shells, crustacea, &c. are in the progress of formation. This specimen, which was dredged up at a few miles from land, is an aggregation of sand with recent marine shells, oysters, muscles, limpets, cockles, &c. with minute corallines; and this example, from off the Isle of Sheppey, consists entirely of cockles *(Cardium edule)*, held together by conglomerated sand. In bays and creeks, bounded by granitic rocks, the bed is found to be composed of micaceous and quartzose sand, consolidated into what may be termed regenerated granite. Off Cape Frio, solid masses of this kind were formed in a few months, and in them were imbedded dollars, and other treasures from the wreck of a vessel, to recover which an exploration by the diving-bell was undertaken.

31. CURRENTS, AND THEIR EFFECTS.—The distribution, over the bottom of the sea, of the detritus brought down by rivers and streams, and of the materials worn away by the action of the waves on the shores, is principally effected by the action of currents, which, from their regularity, permanency, and extent, may be considered as the rivers of the ocean. To this agency I can but briefly allude, and will only instance the Gulf-stream, which is the great current that transports the waters, and the temperature of the tropical regions, into the climates of the north. From the mouth of the Red Sea a current about 50 leagues in breadth sets continually towards the south west; doubling the Cape of Good Hope, it assumes a north-west direction, and in the parallel of St. Helena its breadth exceeds 1000 miles; then taking a direction nearly east, it meets in the parallel of 3° north, along the northern coast of Africa, with a current from the north; entering the Gulf of Florida, they are reflected and form the Gulf-stream, which, passing along the coast of North America, stretches across the Atlantic to the British Isles. At the parallel of 38°, nearly 1000 miles from the Straits of Bahama, the water of the stream is ten degrees warmer than the air. The course of the Gulf-stream is so fixed and regular, that nuts and plants from the West Indies are annually thrown ashore on the western islands of Scotland. The mast of a man-of-

war, burnt at Jamaica, was driven ashore several
months afterwards on the Hebrides, "after perform-
ing a voyage of more than 4000 miles under the
direction of a current which, in the midst of the
ocean, maintains its course as steadily as a river
upon the land."* The transportation of detritus,
resulting from the action of such a current, is obvious;
and we therefore need not wonder at finding the
productions of one country, so frequently included
among the fossils of another.

32. TUFA, INCRUSTING SPRINGS, &c.—The phenomena hitherto considered, are referable to the
mechanical action of water; and the process has been
one of disintegration, and destruction, in the first
instance ; and in the second, of accumulations of sediments in water-channels, and in the bed of the sea.
We must now refer to an operation of a totally different character—the power possessed by streams,
as clear and sparkling as poet ever feigned, or sung,
of consolidating loose materials, of converting porous
strata into solid stone, and of filling up their own
channels by the deposition of calcareous matter.

That most fresh-water holds a greater or less proportion of carbonate of lime in solution, is well
known ; and also that change of temperature, as
well as many other causes, will occasion the calcareous
earth to be in part or wholly precipitated. The *fur*,
as it is called, that lines a boiler which has been long

* Playfair's Works, edition 1822; vol. i. p. 414.

in use, affords a familiar illustration of this agency. At the temperature of 60°, lime is fusible in 700 times its weight of water; and if to this solution a small portion of carbonic acid be added, a carbonate of lime is formed, and precipitated in an insoluble state. If, however, the carbonic acid be in such quantity as to supersaturate the lime, it is again rendered soluble in water; and it is thus that carbonate of lime, held in solution by an excess of fixed air, not in actual combination with the lime, but contained in the water and acting as a menstruum, is commonly found in all waters. An absorption of carbonic acid, or a loss of that portion which exists in excess, will therefore occasion the calcareous earth to be set free, and precipitated on any substance in the water, such as stones, sprigs and leaves of trees, &c. Some springs contain so large a proportion of calcareous earth when they first issue from the rocks, and so speedily throw it down in their course, that advantage has been taken of this circumstance to obtain incrustations of various objects, as leaves, branches, baskets, nests with eggs, and even old wigs. The incrusting springs in Derbyshire are celebrated for such productions. These depositions are termed tufa, or travertine; and in Italy, and many other countries, they constitute extensive beds of concretionary limestone, which is often of a crystalline structure. Even the Cyclopean walls and temples of Pæstum, are formed of this aqueous deposite. At the baths of San Filippo, in

E

Tuscany, where the waters are highly charged with
tufa, this property is applied to a very ingenious
purpose.   The stream is directed against moulds of
medallions, and other bas-reliefs, and very beautiful
casts are thus obtained; of which we have an example in this medallion, which bears the head of
Napoleon, and was presented to me by the Marquis
of Northampton.

33. INCRUSTATIONS NOT PETRIFACTIONS.—AS
incrustations of this kind are commonly, but errone-
ously, termed petrifactions, I will briefly explain
their real nature.   We have before us several
incrustations from various places : baskets of shells,
and nests with eggs, from Derbyshire, for which I
am indebted to the kindness of Sir George Sitwell,
Bart.; a bird, from Knaresborough, Yorkshire, presented by Mr. Thorby; and a branch, partially incrusted, from Ireland, by Miss Ellen Mahony.

TAB. 2.   INCRUSTATION.

I need scarcely observe, that on breaking such
specimens, we find the substances enclosed to have
undergone no change but that of decay, in a greater
or less degree.   In this incrusted bird's nest, the

twigs of which it is composed, like the branch above mentioned, are exposed in several places, and, as you perceive, are not permeated by stony matter, but are dry, and brittle. Now, a true *petrifaction* is altogether of a different nature : the substance being saturated throughout with mineral matter; if we break it, we find that every part of its structure has undergone a change ; sometimes flint has filled up every interstice, and upon slicing and polishing it, the most delicate texture of the original may be detected. Wood, for instance, which is so commonly petrified by flint or chalcedony, may be cut so thin, that with a powerful lens the ramifications of the vessels and the structure of their tissues may be seen, and from their form, and disposition, we may determine the particular kind of tree to which the specimen belonged, although it may, during countless ages, have been cased up in stone. When bone is petrified, the same phenomena are observable ; the most delicate parts of the internal structure are preserved, and all the cells filled up with stone or spar, oftentimes of a different colour from that of the walls of the cells, thus forming a natural, anatomical preparation, of great beauty, and interest.

34. LAKE OF THE SOLFATARA.—This celebrated lake lies in the Campagna between Rome and Tivoli, and is fed by a stream of thermal water which flows into it from a neighbouring pool. The water is of a high temperature, and is saturated with carbonic acid gas, which, as the water cools, is constantly

escaping, and keeping up an ebullition on the surface. The formation of travertine is so rapid, that not only the vegetables, and shell-fish, are surrounded and destroyed by the calcareous deposition, but insects also are frequently incrusted. In these beautiful specimens of travertine from Solfatara, for which I am indebted to Dr. Jenks, vegetable impressions are distinctly seen; and the cavities in the mass, have evidently been occasioned by the decomposition of the vegetable matter.* The stream that flows out of the lake fills a canal, which is conspicuous at a distance, from the line of vapour emanating from the water.

A considerable number of the edifices of both ancient and modern Rome, are built of travertine, derived from the quarries of Ponte Luccano, which have clearly originated from a lake of the same kind. Pæstum is also built of calcareous tufa, derived from similar deposites formed by lakes. "The waters of these lakes," says Sir Humphry Davy, "have their rise at the foot of the Appennines, and hold in solution carbonic acid, which has dissolved a portion of the calcareous rocks through which it has passed; the carbonic acid is dissipated by the atmosphere, and the marble, slowly precipitated, assumes a crystalline form, and produces coherent stones. The acid originates in the action of volcanic fires on the calcareous rocks of which the Appennines are com-

* See Appendix B.

posed; and carbonic acid being thus evolved, rises
to the source of the springs derived from the action
of the atmosphere, gives them their impregnation,
and enables them to dissolve calcareous matter."

35. MARBLE OF TABREEZ.—In Persia, a beautiful
transparent limestone, called Tabreez marble, is
formed by deposition from a celebrated spring near
Maragha, where the whole process of its formation
may be seen. In one part the water is clear, in
another dark, muddy, and stagnant; in a third it is
very thick, and almost black; while in the last stage
it is of a snowy whiteness. The petrifying pools
look like frozen water: a stone thrown on them
breaks the crust, and the water exudes through the
opening. In some states the petrifying process has
proceeded so far as to admit of being walked upon.
A section of the stony mass resembles an accumulation of sheets of paper, being finely laminated.
Such is the tendency of the water to solidify, that
the very bubbles on its surface become hard, as if
they had been suddenly arrested, and metamorphosed
into stone.*

36. STALACTITES, AND STALAGMITES.—By the
infiltration of water through limestone rocks, into
fissures and cavities, sparry concretions are produced
on the roofs, sides, and floors of caverns. The concretionary masses which are dependant from the
roof, resembling icicles, are called stalactites; and

* Morier's Travels.

those which form on the floor, from the droppings
of the water, are termed stalagmites ; and when, as
frequently happens, the two unite, a singularly picturesque effect is produced,—the caves appearing
as if supported by pillars of the most extraordinary
variety and beauty.* Sometimes a linear fissure in
the roof, by the direction it gives to the dropping
of the lapidifying water, forms a transparent curtain
or partition. A remarkable instance of this kind
occurs in a celebrated cavern in North America,
called Weyer's Cave, which is situated in the limestone range of the Blue Mountains.† There are
many caverns in England, celebrated for the variety
and beauty of their sparry ornaments : those in
Derbyshire are well known.

37. GROTTO OF ANTIPAROS.—The Grotto of
Antiparos in the Grecian Archipelago, not far from
Paros, is justly admired. The sides and roofs of its
principal cavity are covered with immense incrustations of calcareous matter, which form either stalactites, depending from above, or irregular pillars
rising from the floor. Several perfect columns
reaching to the ceiling have been formed, and others
are still in the course of formation, by the union of
the stalactite from above, with the stalagmite below.
These being composed of matter slowly deposited,
have assumed the most fantastic shapes ; while the
pure, white, and glittering spar, beautifully catches

* Appendix C.   † Appendix D.

and reflects the light of the torches of the visitors
to this subterranean palace, in a manner which
causes all astonishment to cease at the romantic
tales told of the place—of its caves of diamonds, and
its ruby walls; the real truth, when deprived of all
exaggeration, being sufficient to excite admiration,
and awe. Some of these concretions form a thin
curtain, which is perfectly transparent.*

The specimens which I have selected from my
collection, to illustrate these remarks, exhibit the
usual character of stalactitical concretions; these
long stony icicles are from Portland; and these
minute straws of spar, from an archway near the
Chain Pier, have been formed by the infiltration of
rain through the superincumbent bed of calcareous
rock. This specimen of pebbles, held together by
calc spar, is from the cliffs at Kemp Town; and
affords a proof that in periods very remote, the
same process was in action along our shores. These
beautiful slabs of marble are portions of stalagmite,
from St. Michael's Cave, Gibraltar; and this large
conical mass, which has been cut through and
polished to show its structure, was dug up on the
summit of Alfriston Hill, in this county, and must
have been formed in some chalk cavern, of which
no traces now remain.

38. Consolidation of Sand, and Loose

* See an interesting Essay on Grottoes, in the Saturday
Magazine.

MATERIALS BY INFILTRATION.—The changes effected in loose strata by this process are, however, of still greater importance; for by an infiltration of crystallized carbonate of lime, sand is converted into sand-stone,—fragments of soft chalk are transmuted into a solid rock, as in the Coombe rock of Brighton,—and accumulations of beach, and gravel, into a hard conglomerate, as in this example of the ancient shingle bed of the cliffs, at Rottingdean,—shells, into a building stone, as in this mass from Florida,—and broken corals, into limestone, as in these specimens from Bermuda. By this agency, the bones of animals become permeated with calcareous spar, and the medullary cavities lined with crystals of carbonate of lime : and clay, which has cracked by drying, has its fissures filled up, and becomes consolidated into those curious masses, called Septaria, which when polished, form the beautiful slabs for which Weymouth is so celebrated.

39. DESTRUCTION OF ROCKS BY CARBONIC ACID GAS.—Although, in the instances cited above, water by its combination with carbonic acid, occasions the solidification of loose and porous beds of detritus, yet the effect of this gas on certain rocks is that of disintegration; for by its solvent influence on the felspar, granite itself is reduced to a friable state; the quartz and mica, which with felspar constitute granite, being set at liberty. Mr. Lyell mentions, that the disintegration of granite, is a striking feature in large districts in Auvergne,

especially in the neighbourhood of Clermont. In the ancient bed of shingle in the cliffs at Kemp Town, blocks of granite occur; and here is an example, which may be crumbled to pieces between the fingers. I have already shown you masses of pebbles held together by calcareous spar, from the same beach; we have, therefore, examples in that ancient beach both of the conservative, and disintegrating effects of carbonic acid—cementing the loose beach into solid blocks by incrustation; and, when in a gaseous state, or combined with water, dissolving the granite by its action on the felspar.

40. CARBONIC ACID GAS IN CAVES AND WELLS. —The escape of carbonic acid through fissures, into wells, and caverns, is of frequent occurrence; and as the specific gravity of this gas is greater than that of atmospheric air, it occupies the bottom of these cavities, and its presence is seldom suspected till shown by its deleterious effects.

A melancholy accident that recently occurred at Petworth, the particulars of which were communicated to me by the Earl of Egremont, arose from the sudden escape of carbonic acid into a well, by which two men, employed in the excavation, were almost instantly suffocated.

The Grotto del Cane, near Puzzuoli, four leagues from Naples, has for centuries been celebrated on account of the carbonic acid gas, which rises from fissures in the rock. The gas being spread over the floor of the cave, like a pool of water, its effects are

not perceptible to a creature whose organs of respiration are placed above the level of this invisible
mephitic lake ; but if a dog, or other small animal,
enter the cave, it instantly falls senseless, and would
expire if not speedily removed ; the name of the
cave is obviously derived from the experiment being
often made on dogs, for the amusement of visitors.*

41. CONSOLIDATION BY IRON.—Water charged
with a large proportion of iron, acts an important
part in the consolidation of loose materials, converting sand into iron-stone, and beach or shingle
into ferruginous conglomerates. In this example
of a horse-shoe firmly impacted in a mass of pebbles and sand, presented by Davies Gilbert, Esq.,
the cement which binds the mass, is derived from
the iron.   Nails are frequently found in the centre
of a nodule of hard sandstone formed by this process ; the nail supplying the water with the materials by which the surrounding sand is changed into
stone.   In this very interesting mass of breccia,†
which has been produced by a like process, are two

* See Sandys' Travels.

† The specimen was dug up ten feet below the bed of the
river Dove, in Derbyshire; and the coins are presumed to be
part of the treasures contained in the military chest of the Earl
of Lancaster, which was lost in crossing the river in the dark;
the guards being alarmed by a sudden panic, and the chest with
all its contents thrown into the Dove.   The Earl of Lancaster
was beheaded in March 1322: the specimen before you was
discovered about six years since, more than five centuries after
its submersion.—*See the Vignette of the Title-page.*

silver pennies of Edward I. ; and this curious specimen, for which I am indebted to G. Grantham, Esq.
of Lewes, was procured from a Dutch vessel,
stranded off Hastings a century ago, and is a conglomerate of glass beads, knives, and sand ; the
cementing material having been derived from the
oxidation of the blades.

42. IRON FROM A MORASS.—These masses of oxide
of iron were dug up in a marshy soil, near Bolney,
in Sussex, and are of the same nature as the substance called bog-iron ore, which occurs in peat.
The ebony colour of the woods from Ireland, which
we have already examined, has been occasioned by
an impregnation of iron.   Specimens of bog-iron
are not uncommon in the superficial loam and
gravel of this part of England.

The consolidation of sand and other loose materials by these agencies, is taking place everywhere ;
on the shores of the Mediterranean ; on the coasts
of the West India Islands ; of the Isle of Ascension ;
and on the borders of the United States ; and thus
the remains of man, at Guadaloupe—of turtles, in
the Isle of Ascension—of recent shells, and bones of
ruminants, at Nice—of ancient pottery, in Greece—
and of vegetables, and other substances, in our own
country, have become imbedded and preserved.

I now proceed to notice a few instances of these
most interesting and important operations, by which
much of the solid crust of the globe is perpetually
being renewed.

43. RECENT FORMATION OF MARINE LIMESTONE
IN THE BERMUDAS.—The valuable series of specimens before me (presented by W. D. Saull, Esq.)
is from the Bermuda Islands, and affords examples
of this class of deposites in different states of formation. On the shores of the Bermudas a most interesting deposition of limestone is taking place which
is principally composed of calcareous materials
thrown up by the sea. The ocean which surrounds
the Bermudas abounds in corals and shells, and
from the action of the waves on the coral reefs, and
on the dead shells, the waters become loaded with
calcareous matter. Much of this detritus, no doubt,
is carried down to the profound depths of the
ocean, and there envelopes the remains of animals
and vegetables, thus forming new strata for the use
of future ages ; but a great proportion is wafted
by the waves towards the shores, and is deposited
in the state of fine sand. This sand is drifted
inland by the winds, and becomes more or less
consolidated by the percolation of water, and the
infiltration of crystallized carbonate of lime ; and a
fine white calcareous stone is thus formed, which in
some localities is sufficiently compact for building.
Imbedded in this limestone are numerous shells, and
corals, of the species which inhabit the neighbouring
seas : in some instances the large mottled trochus,
so well known to collectors both in its natural and
polished state, with all its colours preserved, is imbedded in a pure, white, limestone; in many, the

colours are faded, and the shell very much in the
state of the fossils found in the tertiary strata at
Grignon—in others the shell is wanting, but the
hard limestone retains its form and markings. The
corals are imbedded in a similar manner; and masses
are seen in the limestone so like the fossil corals of
the oolite of this country, that it requires an experienced eye to detect their real character.

These specimens, as you perceive, show the transition from loose sand to a solid rock. We have—

1. Broken shells and corals, retaining their colours.

2. Similar materials, more comminuted and completely blanched.

3. An aggregation of fine sand, broken shells,
and corals.

4. Coarse friable limestone, resembling soft chalk,
and composed of comminuted corals, &c.

5. Hard limestone, of similar materials.

6. Compact limestone, enclosing shells, and pebbles.

7. A fine indurated limestone, so hard as to be
with difficulty broken by the hammer, enclosing
a few shells, and corals : this stone is employed
for building.

Mr. Lyell, some years since, in a valuable paper
in the Transactions of the Geological Society,
showed that in the lakes of Scotland a fresh-water
limestone, imbedding the remains of recent shells,
and aquatic plants, was in a state of formation. In
this interesting series of specimens from Forfarshire,

for which I am indebted to the liberality of my
distinguished friend, there are various species of
fresh-water shells, and masses of that common lacustrine plant, the *Chara medicaginula*, beautifully preserved ; even the minute seed-vessels of the chara
are found converted into stone, in precisely the same
manner as is observable in the ancient fresh-water
tertiary limestones. And in the formations at Bermuda, we have an instance, that the sea is, at this time,
producing effects analogous to those which have given
rise to many of the secondary rocks of Europe.

   44. FOSSIL HUMAN SKELETON OF GUADALOUPE.
—Similar formations are in progress along the shores
of the whole West Indian Archipelago; and in St.
Domingo they have greatly extended the plain of
Cayes, where accumulations of conglomerates occur,
and in which, at the depth of 20 feet, fragments of
ancient pottery have been discovered. On the
north-east coast of the main land of Guadaloupe, a
bed of recent limestone forms a sloping bank, or
glacis, from the steep cliffs of the island to the sea,
and is nearly all submerged at high tides. This
modern rock is composed of consolidated sand and
comminuted shells and corals, of species now inhabiting the adjacent seas. Land shells, fragments of
pottery, stone arrow-heads, carved stone and wooden
ornaments, and human skeletons, are found therein
enveloped. This being the only known undoubted
instance of the occurrence of human bones in solid
limestone, has excited great attention ; and the fact,

simple and self-evident as is its history, has been
made the foundation of many vague, and absurd
hypotheses.

In most instances the bones are dispersed; but a
large slab of rock, in which a considerable portion of
the skeleton of a female is imbedded, is preserved
in the British Museum, and has been described by
Mr. König, in a highly interesting memoir in the
Philosophical Transactions, of 1814.

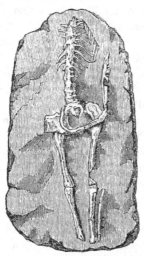

TAB. 3.   FOSSIL HUMAN SKELETON, FROM GUADALOUPE.

The annexed representation will serve to convey
an idea of this celebrated relic, which was detached
from the rock at the Mole, near Point-a-Pitre.

In this specimen the skull is wanting, but the spinal column, many of the ribs, the bones of the left arm and hand, of the pelvis, and of the thighs and legs, remain. The bones still contain some animal matter, and the whole of their phosphate of lime. It is not a little curious, that the fragments of the skull of this very specimen have, during the present year, been described by Professor Moultrie, of the Medical College of South Carolina; having been purchased for the Museum of that State, of a French naturalist, who brought them from Guadaloupe. These relics consist of portions of the temporal, parietal, frontal, sphenoidal, and inferior maxillary bones, of the right side of the skull. An entire skeleton was also discovered in the usual position of burial; another, which was in a softer sandstone, was in a sitting posture. The bodies, thus differently interred, may have belonged to distinct tribes. General Ernouf, who carefully investigated this interesting deposite, conjectured that the occurrence of the scattered bones might be explained by the tradition of a battle, and massacre of a tribe of Gallibis, by the Caribs, on this spot, about 120 years ago; the skeletons being covered by sanddrift from the sea, which became converted into stone. Dr. Moultrie, however, from a rigorous examination, and comparison of the bones of the skull in his possession, is of opinion, that the specimen in the British Museum did not belong to an individual of the Carib, but to one of the Peruvian race, or of a tribe possessing a similar craniological development. In another fossil human skeleton from Guadaloupe, now in the museum of the *Jardin des Plantes*, and represented in the last edition of Cuvier's *Théorie de la Terre*, the figure is bent, the spine forms an arc, and the thigh is drawn up as if the individual were in a sitting posture; a portion of the upper jaw, and the left half of the lower, with several teeth, nearly the whole of one side of the trunk and pelvis, and a considerable portion of the upper and lower left extremities, are preserved. The stone is a travertine, and encloses terrestrial, and marine shells; it is evident that the former have been drifted by the streams from the interior, and the latter deposited by the ocean. In the bed from which the block was extracted have been found teeth of the caiman (crocodile), stone hatchets, and a piece of wood, having rudely sculptured on one side a mask, and on the other the figure of an enormous frog: it is of guaiacum wood, but has become extremely hard, and as black as jet.

45. IMPRESSIONS OF HUMAN FEET IN SANDSTONE. — In connexion with the occurrence of human bones in limestone, I will here notice a discovery of the highest interest, but which has not, as yet, excited among scientific observers the attention which its importance demands. I allude to the fact announced in the American Journal of Science, vol. v. for 1822, of impressions of human feet in sandstone, observed many years ago in a quarry at St.

Louis, on the western bank of the Mississippi. The

TAB. 4. PRINTS OF HUMAN FEET IN SANDSTONE.

above figure is an exact copy of the original drawing, and exhibits the impression of the soles of two corresponding human feet, placed at a short distance from each other, as of an individual standing upright, in an easy position. The prints are described as presenting the perfect impress of the feet, and toes ; exhibiting the form of the muscles, and the flexures of the skin, as if an accurate cast had been taken in a soft substance. They were at first supposed to have been cut in the stone by the native Indians, but a little reflection sufficed to show that they were beyond the efforts of these rude children of nature ; since they evinced a skill, and fidelity of execution, which even my distinguished friend, Sir Francis Chantrey, could not have surpassed. No doubt exists in my

mind, that these are the actual prints of human feet
in soft sand, which was quickly converted into solid
rock by the infiltration of calcareous matter, in the
manner already described. The length of each foot
is ten inches and a half, the spread of the toes four
inches, indicating the usual stature; and the nature
of the impression shows that the feet were uncon-
fined by shoes, or sandals. This phenomenon, unique
of its kind, is fraught with so much importance, that
I have requested Professor Silliman to ascertain the
nature of the sandstone, and the period of its formation. Hereafter I shall have to direct your attention
to impressions of another kind, in rocks of immense
geological antiquity.

46. ISLE OF ASCENSION.—This island, a volcanic
cone in the midst of the Atlantic, which appears to
have been a dome of trachytic rocks, subsequently
affording vent to lava currents, has its shores bounded
by a conglomerate of sand with comminuted shells,
corals, echini, and fragments of lava. In the suite
of specimens before you, presented me by Mr. Lyell,
are portions of this conglomerate in various states
of consolidation. The specimens are composed of
corals, which still retain their colour; shells, more or
less broken; sand formed of similar materials; and
pebbles of trachytic and glassy lavas. The shores
of this island are a favourite resort of turtles, which
repair thither in immense numbers, and deposit their
eggs in the loose sand : the rapid conversion of the
coarse, calcareous banks into solid stone, occasions

the frequent imbedding, and preservation of the eggs;
and there are specimens in the cabinet of the Geological Society, in which the bones of young turtles,
just on the point of being hatched, are well preserved.*
The conglomerate of the Isle of Ascension is, as you
may observe, principally composed of corals. Here
we have another example of a rock formed of the
calcareous skeletons of those wonderful forms of
organic existence. It is not my intention in this
place to dwell on the geological changes produced by
recent zoophytes, in the formation of coral reefs, &c.;
as the examination of the recent, and fossil corals,
will form the subject of a distinct lecture.

47. DRIFTED SAND.—We have already briefly
alluded to the encroachments on the land by the
drifting of sand-banks, thrown up beyond the reach
of the tide, and borne by the winds inland; thus
effecting the desolation of whole regions by their
slow, but certain progress. Egypt instantly presents
herself to the imagination, with her stupendous
pyramids, the sepulchres of a mighty race of
monarchs, and the wonder of the world—her temples, and palaces, once so splendid and massive, as to
bid defiance to the ravages of time—her plains, and
valleys, once teeming with abundance, and supporting
a numerous population—now stripped of her ancient
glories, her fairest regions depopulated, and converted into arid wastes,—her cities overwhelmed,
and prostrate in the dust—and the colossal monuments

* See Lyell's Principles of Geology, 5th edit. vol. iii. p. 269.

of her kings, and the temples of her gods, half buried
beneath the sands of the desert ! And this melancholy instance of the vanity of human power, and
magnificence, is the result of that simple agency,
which this affecting desolation serves so strikingly
to illustrate. The drifting of the sands of the Lybian
desert by the westerly winds, has left no lands capable
of cultivation on any part of the western bank of the
Nile not sheltered by mountains; and in Upper
Egypt, whole districts are covered by moveable
sands, and here and there may be seen the summits
of temples, and the ruins of cities which they have
overwhelmed. "Nothing can be more melancholy,"
says Denon, "than to walk over villages swallowed up
by the sand of the Desert, to trample under foot their
roofs, and minarets, and to reflect that yonder were
cultivated fields—that there grew trees—that here
were the dwellings of men, and that all have now
vanished." The sands of the Desert were in ancient
times remote from Egypt; and the Oases which
still appear in the midst of this sterile region, are
the remains of fertile soils which formerly extended
to the Nile.*

48. SAND-FLOOD, AND RECENT LIMESTONE OF
CORNWALL.—On many parts of the shores of Scotland, sand-floods have converted tracts of great fertility into barren wastes; and on the northern coast
of Cornwall an extensive district has been covered

* See an Essay on the Moving Sands of Africa, in Professor
Jameson's Cuvier's Theory of the Earth, p. 375.

by drifted sand, which in some places forms ranges
of low mounds, or hills, forty feet high, and has become consolidated by the percolation of water holding
iron in solution. This sandstone offers a striking
and most interesting example of recent formation,
and has been described by Dr. Paris, in a memoir
which I do not hesitate to characterize as one of the
most graphic, and instructive geological essays on
modern deposites, that has appeared in this country.*
The sand has clearly been originally drifted from
the sea by hurricanes, probably at a remote period,
and is first seen in a slight, but increasing state of
aggregation, on several parts of the shore in the Bay
of St. Ives. Around the promontory of New Kaye,
the sandstone occurs in various states of induration,
from that of a friable aggregate, to a stone so compact, as to be broken with difficulty by the hammer;
and which is used in the construction of churches,
and houses. Upon examining the stone with a lens,
it appears to be principally made up of comminuted
shells; and it is worthy of remark, that the shelly
particles are frequently found to be spherical, from
the previous operation of water, and much resemble the ancient limestone called oolite, which will
hereafter come under our notice. The rocks upon
which the sandstone reposes are clay, slate, and slaty
limestone; and the water effecting their decomposition may have thus obtained the iron, alumina, and

Appendix E.

other mineral matters by which the loose sand has been converted into sandstone.

The infiltration of water thus impregnated, Dr. Paris observes, is a common and extensive cause of lapidification : at Pendean cove, granitic sand is gradually hardening into breccia, by this process ; and in the island of St. Mary, Dr. Boase has noticed granitic sand becoming indurated by the slow action of water impregnated with iron.

49. SILICEOUS DEPOSITIONS.—Siliceous earth, or the earth of flint, is another abundant mineral, and constitutes so large a proportion of the rocks and strata, that it is computed to form, either in a pure or combined state, nearly one-half of the solid crust of the globe. The flints from our cliffs, the boulders and gravel on our shores, and the pebbles of agate, quartz, and chalcedony, are well-known examples of the usual varieties of silex.*

* Here I would digress for a few moments, to notice an opinion which is so generally prevalent, that I may be permitted to assume, that even some of my auditors may not be prepared at once to answer the question,—Do stones grow? The farmer who annually ploughs the same land, and every year observes a fresh CROP of stones, would probably answer in the affirmative ; and the general observer, who had for successive years noticed his gardens and plantations strewed with stones, notwithstanding their almost daily removal, might entertain the same opinion. A moment's reflection, however, will serve to show, that it is impossible stones can be said to grow, in the proper acceptation of the term. Organic bodies grow, because they are provided with vessels by which they are capable of taking up and assimilating particles of matter, and converting them into their own substance ; but an inorganic body can only increase in bulk by

I scarcely need observe, that this nodule of flint,
derived from a neighbouring chalk quarry, has once
been in a soft or fluid state; for here we perceive
impressions of shells, and of the spines of an echinus
deeply imprinted on its surface.* We have already
seen that water impregnated with carbonic acid gas
is capable of holding lime in solution; and that beds
of travertine, of limestone, and other calcareous
deposites, have originated from this agency; and
although, even in the present advanced state of
chemical knowledge, we are unacquainted with the
process by which any large proportion of silex can
be held in solution by water, yet we have unquestionable proofs, that the solution of siliceous earth
has been effected by natural processes, on a very
extensive scale. At the present moment, Nature,
in her secret laboratories, is still carrying on a
modification of the same process; and of this fact
we have a remarkable instance in the springs of the
Geysers, in Iceland, and of Carlsbad, in Bohemia.
Professor Silliman remarks, that "the sulphuret of
silicon, which is the base of silex, is very soluble,
and that siliceous earth itself is taken up by fixed
alkalies, and by fluoric acid; and these agencies,
like most of those which are chemical, are rendered
more active by heat." A high temperature therefore

the addition of extraneous matter to its outer surface; hence
stones may be incrusted, or they may become conglomerated
together, but they cannot grow.
* *Vide* "Thoughts on a Pebble."

appears necessary to enable water to dissolve a large proportion of silex, &c.; hence, we find that the thermal springs of volcanic regions are the principal agents by which siliceous depositions and incrustations, are at present effected. The Geysers, or boiling fountains, of Iceland, have long been celebrated for possessing this property in an extraordinary degree; holding silex in solution in a large proportion, and depositing it, when cooling, on vegetables and other substances, in a manner similar to that in which carbonate of lime is precipitated by the incrusting springs of which we have already spoken.

50. THE GEYSERS.—Iceland may be considered as a mass of volcanic matter; the only substances not of volcanic origin in the whole island, being beds of surturbrand or bituminous wood, in which occur leaves, trunks, and branches of trees, with clay and ferruginous earth. These strata support an alternation of basalt, tufa, and lava, forming the summit of the hill in which these vegetable remains occur. The Geysers, of which there are a considerable number, are springs, or rather intermittent fountains of hot water, which issue from crevices in a bed of lava. A fountain of boiling water first appears, and is ejected to a considerable height, accompanied with a great evolution of vapour; a volume of steam succeeds, and is thrown up with great force, and a terrific noise like that produced by the escape of steam from the boiler of an engine,

and this operation continues sometimes for more than an hour; an interval of repose of uncertain duration succeeds, after which the same phenomena are repeated. If stones are thrown into the mouth of the cavity, from which the fountain has issued, the stones, after a short interval, are ejected with violence; and again a jet of boiling water, vapour, and steam, appear in succession. Sir G. S. Mackenzie, in his interesting work, "Travels in Iceland," has proposed an ingenious theory, also adopted by Mr. Lyell, to explain these phenomena; and which the following diagram will serve to illustrate.

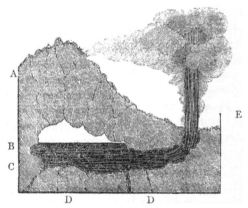

TAB. 5.   PLAN OF THE GEYSERS.

The water from the surface percolates through crevices (A) into a cavity in the rock (B), and heated steam, produced by volcanic agency, rises through the fissures in the lava (D.D). The steam becomes in part condensed, and the water filling the lower part of the cavity (C) is raised to a boiling temperature, while steam under high pressure occupies the upper part of the chasm. The expansive force of the steam becomes gradually augmented, till at length the water is driven up the fissure or pipe (E), and a boiling fountain with an escape of vapour is produced, and continues playing till all the water in the reservoir is expended, and the steam itself escapes with great violence till the supply is exhausted.

The siliceous concretions formed by these springs cover an extent of four leagues; these are specimens of the more friable varieties, presented me by Professor Babbage. M. Eugene Robert, who has recently visited Iceland, states that this curious siliceous formation may be seen, passing by insensible gradations, from a loose friable state, the result of a rapid deposition, to the most compact and transparent marbles, in which impressions of the leaves of the birch-tree, and portions of the stem are distinctly perceptible, and present the appearance of the agatized woods of the West Indies. Stems and leaves of Equiseta, and different mosses, also occur, but none of these plants now exist in the island, the species appearing to have been wholly destroyed by the siliceous deposites. Numerous thermal springs, in the midst of which the Geysers are situated, occupy the valley in the interior of the island. It is evident

that these waters arise from deep crevices in which
they have been heated by volcanic fires. The rivers
proceeding from the springs often resemble milk in
appearance, owing to the argillaceous bole which
they take up in their passage among the siliceous
concretions : such are the white rivers of Olassai.
Mount Hecla, like all the mountains of Iceland, is
entirely covered with snow, and no smoke appears
on its summit. Accumulations of rolled masses of
obsidian and pumice-stone form a layer on the flanks
of the mountain, thirty feet thick ; fragments of
branches of the birch-tree occur in the midst of this
bed ; they are the remains of the ancient forests of
the island, which the volcanic eruptions have entirely extirpated.*

This extensive modern formation of siliceous deposites, is a fact of great interest and importance.
It tells us in language that cannot be mistaken, that
the most solid and refractory substances may be reduced into a liquid state, and assume other modifications, merely by the agency of thermal waters ;
hence the envelopment of the delicate corals, shells,
and spines, in flint nodules, is readily explained.

51. HERTFORDSHIRE BRECCIA, OR PUDDING-
STONE.—We have before us a collection of conglomerates formed by carbonate of lime ; in other words,
an aggregation of pebbles, sand, shells, and corals,
cemented together by calcareous spar and by ferruginous solutions : but this specimen is an example

* Bulletin de la Société Géologique de France.

of a mass of rounded flint pebbles imbedded in a siliceous paste, forming the well-known substance called Hertfordshire puddingstone, which was formerly in great request, for the siliceous cement, being as hard and solid as the pebbles themselves ; the stones may be cut and polished by the lapidary into a great diversity of ornaments. The formation of this breccia must have been effected by a stream of siliceous matter injected into a bed of gravel, converting some masses of the loose pebbles into a solid rock, while those parts, which the melted flint did not reach, remained a layer of loose water-worn materials. It is not my intention in this lecture to dwell on the silicification* of the remains of animals and plants ; I will only remark, that in the deposites of the Geysers, delicate vegetable substances are beautifully preserved ; and that in these silicified woods from the West Indies the most minute structure may be detected, although the specimens will strike fire with steel.

52. EFFECTS OF HIGH TEMPERATURE.—The phenomena presented to our notice in this investigation of the Geysers of Iceland, lead to the consideration of another agent in the transmutations that take place in the crust of the globe. It must be obvious to any intelligent mind, that beds of unconnected and porous materials can have acquired hardness and solidity only by one of the following processes,

---

* Petrifaction by flint.

namely :—1st, by matter dissolved in a fluid, and subsequently deposited among the porous mass in the manner just described; or, 2dly, by their reduction by heat into a state of softness or fusion, and afterwards cooling into a solid mass.* Fire—or to speak more correctly, high temperature, however induced, whether by electro-magnetic influence, or from central or medial source of heat—and water, are therefore the great agents by which the condition of the surface of our planet is modified. We have already seen how vast are the changes which result from the effects of the latter; we must now take a rapid survey of the influence which the former is capable of exerting; an influence far more universal, and varied, than we may at first be prepared to expect. The expansive power of heat on most substances, its conversion of the most solid and durable bodies, first into a fluid, and lastly into a gaseous state, are phenomena so familiar as to require no lengthened comment. But the effects of heat are found to vary according to the circumstances under which bodies are submitted to its operation, and hence the changes induced by high temperature under great pressure, are totally different from those effected by fire on the surface, under the ordinary weight of the atmosphere. A familiar example will best illustrate my meaning. Chalk consists of lime combined with carbonic acid; and as for agricultural, and other economical purposes, it is desirable

* Playfair.

to have the lime in its pure state, the chalk, or
limestone, is exposed to a great heat, in kilns erected
in the open air, until all the carbonic acid is dissipated, and the chalk is said to be burnt into quick-
lime. In these specimens, you see the same substances in the state both of chalk and lime. Now
it may readily be conceived, that if this operation
were conducted under such a degree of pressure
that the gas could not escape, the formation of
quick-lime would not take place; the chalk would
be fused ; the carbonic acid, released from its
present relation with the calcareous particles, would
enter into other combinations, and the mass when
cooled would be wholly different from the product
of the lime-kilns, formed by the same agency in the
open air. Experiments have proved that this opinion
is correct. Sir James Hall exposed pounded chalk to
intense heat, under great pressure, and it was fused,
not into lime, but crystalline marble : and shells
enclosed in the chalk underwent the same transmutation, yet preserved their forms. That analogous
changes have been effected by a similar operation
in nature, we have abundant proof; but in this
stage of our inquiry it is only necessary to remark,
that where ancient streams of lava have traversed
chalk, the latter invariably possesses a crystalline
structure. We shall hereafter find, in accordance
with the beautiful and philosophical theory of
Dr. Hutton, that all the strata have been more or
less modified by heat, acting under great pressure

and at various depths; and that the present position and direction of the materials composing the crust of the globe, have been produced by the same agency.* The Huttonian theory, indeed, offers a most satisfactory explanation of a great proportion of geological phenomena, enabling us to solve many of the most difficult problems in the science; and it is but an act of justice to the memory of an illustrious philosopher, and of his able illustrator, Professor Playfair, to state that this theory, corrected and elucidated by the light which modern discoveries have shed upon the physical history of our planet, is now embraced by the most distinguished geologists.

53. VOLCANIC AGENCY.—Of the activity and power of the agent to which these remarks more immediately refer, the streams of lava ejected through crevices and fissures of the earth, accompanied with evolutions of heat, and smoke, and vapour, afford the most striking proofs; and the volcano, with its frequent concomitant the earthquake, has in all ages excited the curiosity of mankind. It would be foreign to the design of this discourse, to enter at large upon the nature and causes of volcanic action. Dr. Daubeny,† Mr. Scrope,‡ and others, have published highly interesting treatises

* See Playfair's Illustrations of the Huttonian Theory, vol. i. p. 33, et seq. Edin. 1822.

† Daubeny's Lectures on Volcanoes, 1826.

‡ Scrope's Considerations on Volcanoes, 1825.

on the subject; and Mr. Lyell has given an admirable sketch of volcanic phenomena.* I will only advert to the increased temperature of the earth in proportion as we penetrate the interior, and the profound depths from which thermal waters take their rise, as tending to support the opinion, that volcanic eruptions are occasioned by electrochemical changes, which are constantly going on in the interior of our globe. We shall hereafter have occasion to demonstrate that dislocation of the strata, and elevation of the bottom of the ocean, and eruptions of melted mineral matter, have taken place from the earliest geological periods within the scope of our enquiries.

The expansive power of heat, even in ordinary circumstances, is very considerable, as is shown by the instrument called a Pyrometer, which illustrates a phenomenon continually presented to our notice, namely, the expansion of a bar of metal by heat, and its contraction, by cooling, into its original dimensions. The thermometrical expansion of solid bodies when effected on an enlarged scale, gives rise to many interesting phenomena; and experiments made with great care by Colonel Totten, on the expansion of granite, marble, and other rocks, by variations of temperature, have shown that the mere expansion, or contraction, of extensive beds of these materials, would account for the elevation and subsidence of considerable

* Principles of Geology.

G

tracts of country, and explain many analogous phenomena.*

54. TEMPLE OF JUPITER SERAPIS.—One of the
most interesting examples of local elevation and
subsidence, apparently resulting from this cause, is
afforded by the celebrated remains of the temple of

TAB. 6.—TEMPLE OF JUPITER SERAPIS.
(*From Mr. Lyell's Principles of Geology.*)

Jupiter Serapis, at Puzzuoli; and which my distinguished friend Mr. Lyell, by selecting as the subject
of the frontispiece of his invaluable work, has for
ever associated with the Principles of Geology.

* American Journal of Science, vol. xxii.

These ruins are situated on the shore of the Bay of Baiæ, and consist of the remains of a large building of a quadrangular form, seventy feet in diameter ; the roof of which was supported by twenty-four granite columns, and twenty-two of marble, each formed of a single stone. Many of the pillars are broken and strewed about the pavement, but three remain standing nearly erect, and on these are inscriptions, not traced by Greek or Roman, but by some of the simplest forms of animal existence, which have here left enduring records of the physical changes that have taken place on these shores, since man erected the temple in honour of his gods. The tallest column is forty-two feet in height ; its surface is smooth and uninjured to an elevation of about twelve feet from the pedestal, where a row of perforations made by a species of marine boring muscle (*Modiola lithophaga*) commences, and extends to the height of nine feet ; above which all traces of their ravages disappear.*
The perforations, many of which still contain shells, are of a pear shape, and are so numerous and deep as to prove unquestionably that the pillars were immersed in sea-water, at the very time when the base and lower portions were protected by rubbish and tufa, and that the upper parts projected above the waters, and consequently were placed beyond the reach of the *lithodomi.* The platform of the temple is now about one foot below high-water

* See Appendix F.
G 2

mark ; and the sea, which is only forty yards distant,
penetrates the intervening soil. The upper part of
the band of perforations is, therefore, now at least
twenty-three feet above the level of the sea ; and yet
it is evident that these columns were once plunged
in salt water for a long period. It is equally clear
that they have since been elevated to a height of
twenty-three feet, still maintaining their erect position, amid the extraordinary changes which they
have undergone, and incontrovertibly proving that
the relative level of the land and sea, on that part
of the coast, has changed more than once since
the christian era ; each movement, both of subsidence and elevation, having exceeded twenty feet.*
Yet there stand those marvellous columns at the
present moment,—

> " Flinging their shadows from on high,
> Like dials, which the wizard Time
> Hath raised to count his ages by !"

Professor Babbage, in a valuable paper on these
phenomena, attributes the tranquil elevation and
depression of the temple, to the contraction and
expansion of the strata on which it is built. The
sources of volcanic action in the surrounding country are, as you know, very numerous ; and a hot
spring still exists on the land-side of the ruins.
The change of level is therefore easily accounted
for, by supposing the temple to have been built on

---

* Pinciples of Geology, vol. ii. p. 268.

the surface of rocks, of a high temperature, which subsequently contracted by slow refrigeration. When this contraction had reached a certain point, a fresh accession of heat from the neighbouring volcano increased the temperature of the strata; which again expanded and raised the ruins to their present level.*

Professor Babbage, in the interesting Essay to which I have alluded, carries out the views embodied in these brief remarks, to explain the elevation of continents and mountain ranges; assuming as the basis of his theory the following facts :—

1st. As we descend below the surface of the earth, the temperature increases.

2dly. Solid rocks expand by being heated, but clay and some other substances contract.

3dly. Rocks and strata of dissimilar characters present a corresponding difference as conductors of caloric.

4thly. The radiation of heat from the earth varies in different parts of its surface; according as it is covered by forests, mountains, deserts, or water.

5thly. Existing atmospheric agents, and other causes, are constantly changing the condition of the surface of the globe.

Thus wherever a sea or lake is filled up by the wearing down of the adjacent lands, new beds are formed, conducting heat much less quickly than the water; while the radiation from the surface of the

* Appendix G.

new land will also be different. Hence, any source
of heat, whether partial or central, which previously
existed below that sea, must heat the strata under-
neath, because they are now protected by a bad
conductor.* They must therefore raise, by their
expansion, the newly-formed deposites above their
former level ;—and thus the bottom of an ocean may
become a continent. The whole expansion, however, resulting from the altered circumstances, may
not take place until *long* after the filling up of the
sea ; in which case its conversion into dry land will
result partly from the accumulation of detritus, and
partly from the elevation of the bottom. As the heat
now penetrates the newly-formed strata, a different
action may be induced ; the beds of clay or sand
may become consolidated, and instead of expanding,
may contract. In this case, either large depressions
will occur within the limits of the new continent, or
after another interval, the new land may again subside, and form a shallow sea. This sea may be
again filled up by a repetition of the same processes
as before ;—and thus alternations of marine and
fresh-water deposites may occur, having interposed
between them the productions of the dry land.†

* Sir John Herschel observes, that this process is precisely
similar to that by which a great coat, in a wintry day, increases
the feeling of warmth ; the flow of heat outwards being obstructed, and the surface of congelation removed to a distance
from the body, by the heat thereby accumulated beneath the new
covering.

† Proceedings of the Geological Society, March 1834.

To review the physical changes which are still taking place around the Bay of Naples would prove highly interesting, and I much regret that time will only permit me to remark, that whole mountains have been elevated on the one hand, and temples and palaces are seen beneath the sea on the other. In our sister island we have also evidence of former changes of a like nature; and which are alluded to by our inimitable lyric poet, in the following beautiful lines :—

On Lough Neagh's bank as the fisherman strays,
   When the clear cold eve's declining,
He sees the round towers of other days,
   In the wave beneath him shining!

Thus shall memory often, in dreams sublime,
   Catch a glimpse of the days that are over;
Thus, sighing, look through the waves of time
   For the long faded glories they cover!

55. ELEVATION OF THE COAST OF CHILI.—One of the most remarkable modern instances of the elevation of an extensive tract of country, is that recorded by Mrs. Calcott, as having been produced by the memorable earthquake which visited Chili in 1822, and continued at short intervals till the end of 1823. The shocks were felt through a space of 1,200 miles, from north to south. At Valparaiso, on the morning of the 20th of November, it appeared that the whole line of coast had been raised above its level; an old wreck of a ship, which could not previously be approached, was now accessible from the land; and beds of scallops were brought

to light, which were not before known to exist.
"When I went to examine the coast," says Mrs.
Calcott, "although it was high-water, I found the
ancient bed of the sea laid bare and dry; with
oysters, muscles, and other shells, adhering to the
rocks on which they grew: the fish being all dead,
and exhaling the most offensive effluvia. It appeared
to me, that there was every reason to believe the
coast had been raised by earthquakes at former
periods, in a similar manner; for there were several
lines of beach, consisting of shingle mixed with
shells, extending in parallel lines to the shore, to the
height of fifty feet above the sea." Part of the
coast thus elevated consists of granite; and subsequent observations have proved that the whole of
the country was raised, from the foot of the Andes
to far out at sea: the supposed area over which
the elevatory movements extended, being about
100,000 square miles; a space equal in extent to
half the kingdom of France. Mrs. Somerville mentions, that a further elevation to a considerable extent
has also taken place along the Chilian coast, in consequence of the violent earthquake of 1835.

56. LIFTED SEA BEACH AT BRIGHTON.—Examples of such changes occur in almost every
part of the world; and there is perhaps no considerable tract of country which does not afford
some proof that similar physical mutations have
taken place in modern times. And although I
cannot point out to you a temple of Serapis on our

shores, yet within a few hundred yards of this place, there is unquestionable evidence that the relative level of land and sea, has undergone great changes within, to speak geologically, a comparatively recent period. The upper part of the cliffs, extending from the commencement of the low range by Shoreham, to Rottingdean, is composed of chalk with rubble, flints slightly rolled, and clay and loam; the whole being clearly an accumulation of water-worn materials, deposited in an estuary or bay of the sea. The base of the cliffs, to the height of a few feet, is composed of the solid chalk strata, which may be seen at low-water extending far out to sea, and covered here and there by shingle and sand. Upon the upper surface of the chalk, and interposed between it and the superincumbent mass just described, is a bed of shingle, composed of rolled chalk, flints, pebbles, and sand, with boulders of granite, porphyry, and other rocks, not now met with on our shores; in fact, an ancient sea beach, which has evidently been formed at some remote epoch, on the Sussex coast, in like manner as the present bed of shingle, which skirts the base of the cliffs, is in the course of formation. Among the pebbles of this ancient beach, are rolled masses of chalk and limestone, which are full of perforations made by boring shells; here are several specimens, which, as you perceive, are similar to those made in the chalk-rock by the recent *pholades* and *mytili*. As I shall have occasion to revert to these cliffs in my next lecture, this brief description will be sufficient for our present purpose.

The following diagram represents a vertical section of the cliffs, as seen in those parts where the inroads of the sea have extended to the chalk strata, and the face of the ancient chalk cliff is exposed, the new deposites being shown in profile.

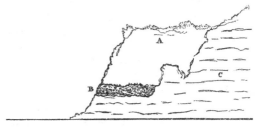

TAB. 7.—ELEVATED BEACH AT BRIGHTON.

(A) Chalk, rubble, loam, &c., obscurely stratified; and called the Elephant-Bed, from its containing teeth and bones of elephants. This constitutes the upper three-fourths of the cliffs.

(B) Shingle, or sea beach and sand, many feet above high-water mark. This ancient shingle, though, from the inroads of the sea, it extends in the cliffs beyond Kemp Town but a short distance inland, is constantly found beneath the loam and clay, several hundred yards from the shore in the western part of Brighton. In a well lately dug in the Western road, the shingle bed occurs at the depth of fifty-four feet.

(C) The undisturbed chalk, which forms a sloping cliff, inland, or behind the beds (A) and (B), passing under the ancient sea beach, and appearing as a terrace at the foot of the present cliffs.*

These appearances demonstrate the following sequence of changes in the relative level of the land and sea on the Sussex shores :—

First. The chalk terrace on which the ancient shingle rests, was on a level with the sea for a long period, and the beach was formed, like the modern beach, by the action of the waves on the then existing cliffs. The rolled condition of the materials, and the borings of the *lithodomi*, prove a change of level as decidedly as do the perforations in the columns of the temple of Serapis.

Secondly. The whole line of coast, with the shingle (B), was submerged to such a depth, as to admit the deposition of the strata (A) above them.

Lastly. The cliffs were raised to their present elevation, and this period was the commencement of the existing sea beach.

The elevation of the sea shore with beds of marine shells, already alluded to as having been produced by earthquakes on the Chilian coast, has here then a parallel ; and, should the correctness of these inferences be questioned, I would beg of you to visit Castle Hill, near Newhaven (about eight miles east of Brighton), and there you will find, immediately

* See Geology of the South-East of England, p. 30; and Fossils of the South Downs, p. 277.

beneath the turf, a regular sea beach and beds of oyster-shells, many feet in thickness, lying on the summit of chalk cliffs, 150 feet above the level of the sea. Near Bromley, in Kent, and at Reading, in Berkshire, similar accumulations of beach and oyster-shells are to be found : specimens from each of these localities are placed on the table before us. Elevated shingles, of comparatively recent epochs, occur on the shores of the Frith of Forth, and also along the western coasts of England ; as my distinguished friend, Mr. Murchison, has satisfactorily demonstrated.

57. ELEVATION OF SCANDINAVIA. — Having thus adduced a few striking proofs of the mutations which the land has undergone in past times, we are led to enquire—Is this change still going on ? Is the alternate subsidence and elevation of the land the effect of a law of nature, established from the existence of the present condition of our planet, and destined to continue in action while its physical constitution remains the same ? We shall hereafter find, that this law has been in constant action from the earliest periods of the earth's history, of which her physical monuments afford any indications ; and I now proceed to adduce an instance in which the elevation of a country, with the whole burthen of its people and its cities, is actually taking place, unheeded by the busy multitude, and known only by the researches of the natural philosopher. I allude to Scandinavia, where it is ascertained, that the

whole country, from Frederickshall, in Sweden, to Abo, in Finland, and even, perhaps, as far as St. Petersburgh, is slowly and visibly rising; while the adjacent coast of Greenland is suffering a gradual depression. The state, therefore, is one of oscillation, the waters appearing to sink at Torneo, and to retain their former level at Copenhagen. The opinion that Sweden is in this state of change is no new idea; it was long since noticed by Celsius,* and other Swedish philosophers. Mr. Lyell has twice visited Scandinavia within the last few years, with the view of determining this interesting question, and has fully convinced himself that certain parts of Sweden are undergoing a gradual rise, to the amount of two or three feet in a century; while other parts, farther to the south, appear to have experienced no movement.† He visited some parts of the shores of the Bothnian Gulf, between Stockholm and Gefle, and of the western coasts of Sweden, districts particularly alluded to by Celsius. He examined the marks cut by the Swedish pilots, under the direction of the Swedish Academy of Sciences, in 1820, and found the level of the Baltic, in calm weather, to be several inches lower than the marks, and several feet below those made seventy or a hundred years ago. Similar results were obtained on the side of the ocean, and in both districts the

* Illustrations of the Huttonian Theory, p. 436, edit. 1822.

† Philosophical Transactions. Principles of Geology, Fifth Edition, vol. ii. p. 286.

testimony of the inhabitants agreed with that of
their ancestors, recorded by Celsius. Mr. Lyell discovered on the shores of the Northern Sea, elevated
banks of recent shells, at various heights, from 10 to
200 feet, and deposites on the side of the Bothnian
Gulf, between Stockholm and Gefle, containing fossil
shells of the same species which now characterize
the brackish waters of that sea. These occur at
various elevations, from one to a hundred feet, and
sometimes reach fifty miles inland. The shells are
partly marine and partly fluviatile ; the marine
species are identical with those now living in the
ocean, but are dwarfish in size, and never attain
the average dimensions of those which live in water
sufficiently salt, to enable them to reach their full
development. The specimens before you were
collected by my valued friend at Uddevalla, in Sweden, from cliffs twenty feet above the level of the
sea ; they consist of recent marine species, such as
inhabit the neighbouring waters.

Of the reality of these changes in the relative
level of the land and the Northern ocean, there cannot exist a doubt ; but the mind is so accustomed
to associate the idea of stability with the land, and
of mutability with the sea, that it may be necessary
to offer a few additional remarks on these highly interesting phenomena. As it is the property of all
fluids to find their own level, it is obvious that if
the level of the sea be elevated or depressed in any
one part, that elevation or depression must extend

over the whole surface of the ocean, and the level therefore cannot be affected by local causes. But movements of the land may take place, and the effect extend over whole countries, as in South America,—or along lines of coast, as in Sussex, — or be confined to a single island, — or even to the broken columns of a temple, as at Puzzuoli.* But while the land is rising in the more northern latitudes, it appears to be sinking on the shores of the Mediterranean. Breislak mentions † that numerous remains of buildings are to be seen in the gulf of Baiæ; ten columns of granite, at the foot of Monte Nuovo, are nearly covered by the sea, as are the ruins of a palace built by Tiberius in the island of Caprea. Thus, while the level of the sea is sinking in the north, it is rising in the Mediterranean; and as all the parts of the ocean communicate, the level of the sea cannot permanently rise in one part and sink in another, but must rise and fall equally to maintain its level. We must therefore consider it as demonstrated, that the relative change of level has proceeded from the elevation or depression of the land, and that these phenomena are produced by the expansive force of heat, or of electro-chemical agency. If we bear in mind the insignificance of the masses affected by these operations, as contrasted with the earth itself, recollecting that the varnish of a small artificial sphere is equal in

* See Playfair's admirable comments on this geological problem.—*Illustrations*, p. 433.   † Playfair.

proportionate thickness to the entire series of strata which the ingenuity of man has been able to explore, we can readily conceive that as fissures and inequalities are produced in that varnish by heat or cold, in like manner the elevation of mountain chains, the rending of countries, and the subsidence of whole continents, may be occasioned by the thermometrical expansion or contraction of the materials of which our planet is constructed.

58. RETROSPECT —In this imperfect review of the geological phenomena, which even a superficial examination of the surface of the globe presents to our notice, I have doubtless dwelt on several subjects which are familiar to many of my auditors. But as one of our most accomplished philosophers has remarked — " The teacher of Geology must suppose himself called on to answer questions both concerning the facts of the science and the inferences to be deduced therefrom ; and his instruction will be so much the more successful as he takes these questions in the most natural order of their occurrence, and answers them most completely and satisfactorily. In doing this he is not at liberty to neglect even elementary truths, for if these were passed over in compliment to such as have made progress in the science ; those for whose advantage he is especially interested, would be called to the unreasonable task of labouring without instruments, and theorizing without intelligible data." *

* Phillips, Guide to Geology.

From the vast field of inquiry over which our observations have extended, it may be useful to offer a brief summary of the leading principles that have been enunciated, and the phenomena on which they are founded. By the most profound and sublime investigations of which the human mind is capable, we learn that our earth is one of countless myriads of spherical bodies, revolving round central luminaries; and that these bodies occur in every variety of condition, from that of diffuse luminous vapour, to opaque solid globes like our own. All the materials of which the earth is composed may exist either in a solid, fluid, or gaseous state; and simply by a change of temperature, or by electro-chemical agency, every substance may undergo a transition from one state to the other. Water existing as ice, fluid, or vapour, and separable into two invisible gases, offers a familiar example of a body constantly exhibiting these changes; and mercury, of a metal which, although generally fluid, or melted, becomes, when exposed to a very low temperature, a solid mass like silver. The relative position of land and water, and the inequalities on the surface of the earth, are subject to constant changes, which are regulated by certain fixed laws. The principal causes of the degradation of the land are atmospheric agencies, changes of temperature, and the action of running water, by which the disintegrated materials of the land are carried into the bed of the ocean. The mud, sand,

H

and other detritus thus produced, are reconsolidated
by certain chemical changes which are in constant
activity, both on the land and in the depths of the
ocean, and new rocks are thus in the progress of
formation. But the conjoint effect of these disintegrating agencies is unremitting destruction of the
lands, and were there no conservative process,
the whole of the dry land would disappear, and the
earth be covered by one vast sheet of water.
The interior of the globe, however, possesses a
source of heat,—and whether this heat exists as a
central nucleus of high temperature, or as medial
foci, — whether dependent on its assumed nebulous state, or produced by electro-magnetic forces
acting on the mineral substances contained in the
interior of our planet,—does not affect the present
inquiry. This internal heat, however produced,
occasions constant changes in the relative level of
the land and water; elevating whole continents,—
converting the bed of the sea into dry land,—and
submerging the dry land into the abyss of the
ocean. The volcano and the earthquake are the
effects of its paroxysmal energies, — the quiet and
insensible elevation of the land, of its slow but
certain operation. By this antagonist power the
accumulation of the spoils of the land, which the
rivers, and waves, and currents have carried into
the bed of the ocean, are again brought to the surface, and form the elements of new islands and
continents; and by the organic remains discovered

in these strata, we trace the nature of the countries from whence these spoils were derived. In the deltas and estuaries of modern times,—in the detritus accumulating in the beds of the ocean,—in the recent tracts of limestone forming on the sea-shores,—in the cooled lava currents erupted from existing volcanoes,—the remains of man and of his works, and of the animals and plants which are his contemporaries, are found imbedded.

Such are the deductions derived from the phenomena which have been submitted to our examination.

To the mind previously uninstructed in geological science, I am ready to acknowledge that to attribute mutability to the rocks and the mountains, must appear as startling and incredible, as did the astronomical doctrines of Galileo to the people of his times. But the intelligent observer, whose attention has been directed to the facts laid before him, even in this brief survey, cannot, I conceive, refuse his assent to the inferences thus cautiously obtained. As we proceed in our investigation, we shall find that from the earliest period of the earth's physical history, its surface has been subject to incessant fluctuation; and as the land has been the theatre of perpetual mutation, that element, which has hitherto been considered as the type of mutability, can alone be regarded as having undergone no change. This idea is finely embodied by Lord Byron in the following sublime apostrophe to the

H 2

Ocean, with which I will conclude this discourse.

> " Thy shores are empires changed in all save thee :
> Assyria, Greece, Rome, Carthage, what are they ?
> Thy waters wasted them while they were free,
> And many a tyrant since—thy shores obey
> The stranger, slave, or savage ;—their decay
> Has dried up realms to deserts ;—not so thou !
> Unchangeable, save to thy wild waves' play ;
> Time writes no wrinkle on thine azure brow—
> Such as Creation's dawn beheld, thou rollest now ! ''

LECTURE II.

1. Introductory Observations.—In the previous lecture we took a comprehensive view of the actual physical condition of the surface of our planet, and the nature and results of the principal agents by which the land is disintegrated and renewed. We found in the modern fluviatile and marine deposites, that the remains of man, of works of art, and of the existing races of animals, were preserved. In every step of our progress, the grand law of nature, alternate decay and renovation, was exemplified in striking characters—whether in the

regions of eternal snow, or in torrid climes—in the
rocks and mountains, or on the verdant plains—by
the agency of heat, or by the effect of cold—of
drought, or of moisture—of steam, or of vapour—
by the abrasion of torrents and rivers—by inundations of the ocean—or by volcanic eruptions—still
the work of destruction, in every varying character,
was apparent. And on the other hand we perceived
that amidst all these processes of decay, and desolation, perpetual renovation was at the same time
going on,—and that Nature was repairing her ruins,
and accumulating fresh materials for new islands,
and continents; and that countless myriads of living
instruments were employed to consolidate, and build
up the rocky fabric of the earth; and that even the
most terrific of physical phenomena, the earthquake,
and the volcano, were but salutary provisions of the
Supreme Cause, by which the harmony and inte-
grity of the earth were maintained and perpetuated.
The occurrence of human skeletons in modern
limestone—of coins and works of art in recent
breccia—and the preservation of the bones of existing species of animals, and of the leaves and
branches of vegetables, in the various deposites that
are in progress, incontestibly prove that enduring
memorials of the present state of animated nature
on our globe, will be transmitted to future ages.
When the beds of the existing seas shall be elevated
above the waters, and covered with woods and
forests — when the deltas of our rivers shall be

converted into fertile tracts, and become the scites
of towns and cities—we cannot doubt that in the
materials, extracted for their edifices, the then existing races of mankind will discover indelible records of the physical history of our times, long
after all traces of those stupendous works, upon
which we vainly attempt to confer immortality, have
disappeared. But we must now proceed, and pass
from the ephemeral productions of man, to the
enduring monuments of nature—from the coins of
brass and silver, to the imperishable medals on
which the past events of the globe are inscribed—
from the mouldering ruins of temples and palaces,
to the examination of the mighty relics, which the
ancient revolutions of the earth have entombed.

2. EXTINCTION OF ANIMALS.—Before entering
upon the examination of the geological phenomena,
which belong to the period immediately antecedent
to the present, it will be necessary to notice one of
the most remarkable facts which geological investigations have established,—namely, the entire obliteration of certain forms of animals and plants.
The fluctuating state of the earth's surface, with
which our previous inquiries have made us familiar,
will have prepared us for the disappearance of some
species of animals;—and here another law of the
Creator is manifest. Certain races of living beings,
suitable to peculiar conditions of the earth, appear
to have been created; and when those states became
no longer favourable for the continuance of such

types of organization, according to the natural laws by which the conditions of their existence were determined, the races disappeared, and were probably succeeded by new varieties of life.

The extinction of whole genera of animals and plants has no doubt depended on many causes. In the earlier ages, the changes of temperature, and the rapid mutations of land and water, were probably the principal agents of destruction; but since man has become the lord of the creation, his necessities and caprice have occasioned the extirpation of whole tribes of animals, whose relics are found in the superficial strata, with those of species concerning which both history and tradition are silent.

In this country the beaver, wolf, hyena, bear, &c. are examples of species which still exist elsewhere; while the Irish elk and the mammoth, whose remains occur in our alluvial deposites, are both extinct; and the former was unquestionably extirpated by the early inhabitants of these islands. The obliteration of certain forms of organization, is therefore clearly dependent on a law in the economy of nature which is still in active operation; and I shall now proceed to notice the connecting links between the actually existing species, and those which are blotted out from the face of the earth.

3. Animals Extinct by Human Agency.— That the extinction of many of the existing races of animals must soon take place, from the immense destruction occasioned by man, cannot admit of

doubt. In those which supply fur, a remarkable proof of this inference is cited in a late number of the American Journal of Science. "Immediately after South Georgia was explored by Captain Cook, in 1771, the Americans commenced carrying seal-skins from thence to China, where they obtained most exorbitant prices. *One million two hundred thousand skins* have been taken from that island alone, since that period; and nearly an equal number from the island of Desolation! The numbers of the fur-seals killed in the South Shetland Isles (S. lat. 63°,) in 1821 and 1822, amounted to three hundred and twenty thousand. This valuable animal is now almost extinct in all these islands." From the most authentic statements it appears certain that the fur trade must henceforward decline, since the advanced state of geographical science shows that no new countries remain to be explored. In North America the animals are slowly decreasing from the persevering efforts, and the indiscriminate slaughter, practised by the hunters, and by the appropriation to the use of man, of those forests and rivers which have once afforded them food and protection. They recede with the aborigines before the tide of civilization.

4. APTERYX AUSTRALIS.—An extraordinary bird, a native of New Zealand, of which no living individual is known, and but one stuffed specimen exists in Europe, appears to be on the point of extinction, if, indeed, it be not already obliterated.

It is the Apteryx represented below, and which de-
rives its name from being destitute of wings.   The
only specimen known to naturalists, was figured and
described by Dr. Shaw, and is now in the collection

TAB. 8.—APTERYX AUSTRALIS.

of Lord Stanley.   It has lately been examined by
Mr. Yarrell, so that the characters of the skeleton are
correctly ascertained.   The creature is of a greyish
brown colour, and has neither wings nor tail.   The
beak is slightly curved, and the nasal apertures, instead of opening at the base, as in birds in general,
and especially in those of a similar conformation of
beak, which is adapted for respiration while immersed in mud or water, is placed at the apex.
The eyes are very small.   The feathers are long
and loose, like those of the emu, but each plume

has a single shaft. The most active inquiries have
not succeeded in obtaining either a living or dead
specimen of this bird, although a missionary informed me that skins of the creature were still
worn as ornaments by the New Zealand chiefs.
There can be no doubt, however, that this wonderful creature either is extinct, or will shortly cease to
exist.

TAB. 9.—THE DODO.

5. THE DODO (*Didus ineptus*).—But lest this al-
leged extirpation of a peculiar type of organization
be considered questionable, let me call your attention to a remarkable instance afforded by the
Dodo, which has been annihilated, and become
a denizen of the fossil kingdom, almost before our
eyes. The Dodo was a bird of the gallinaceous
tribe, larger than a turkey, which existed in great

numbers in the Mauritius and adjacent islands,
when those parts were first colonized by the Dutch,
about two centuries ago. This bird formed the
principal food of the inhabitants, but it was found
to be incapable of domestication, and its numbers
soon became sensibly diminished. Stuffed specimens were preserved in the museums of Europe, and
paintings of the living animal were executed, and are
still extant in the Ashmolean Museum at Oxford, and
in the British Museum. But the Dodo is now extinct
—it is no longer to be found in the isles where it once
flourished; and even all the stuffed specimens are destroyed. The only relics that remain, are the head
and foot of one individual in the Ashmolean, and
the leg of another in the British Museum. To
render this illustration complete, the bones of the
Dodo have been found in a tufaceous deposit, beneath a bed of lava, in the Isle of France; so that
if the remains of the recent bird already alluded to,
had not been preserved, these fossil relics would
have constituted the only record that such a creature had ever existed on our planet.

6. The Irish Elk, or Cervus megaceros.
(*Elk with great antlers.*)—The shell marls of Ireland
also afford evidence of the existence of a creature,
which, like the Dodo, was once cotemporaneous
with the human species, but is now altogether
extinct, the last individual of the race having, in
all probability, been destroyed by man. These
remains commonly occur in the beds of marl

beneath the peat-bogs, which are probably, like those
of Scotland, the scites of ancient lakes. In Curragh,
immense quantities of the bones of the Elk lie within
a small space, as if the animals had assembled in a
herd : the skeletons appear to be entire, and the
nose is elevated, and the antlers thrown back on
the shoulders, seeming to denote that the creatures had sunk in a morass, and been suffocated.
Remains of the Elk occur also in beds of marl and
gravel, in many parts of England, France, Germany, and Italy. This enormous ruminant very
far exceeded in magnitude any living species. The
skeleton is upwards of ten feet high from the ground
to the highest point of the horns; and the antlers,
which are palmated, are from ten to fourteen feet
from one extremity to the other. The museum of
the late eminent anatomist, Joshua Brookes, which,
to the disgrace of the government of this country,
was suffered to be dispersed, contained a magnificent
pair, measuring eleven feet in expanse, which are
now in my collection. Skulls have been found
without horns, and these probably belonged to fe-
males. The average weight of the head and horns
is computed at three-quarters of a hundred weight.
The horns are generally in a fine state of preservation, coloured of a dark brown, with here and there
a bluish incrustation of phosphate of iron, like those
of the deer from Lewes Levels. The Elk shed
its horns, and probably, like existing species, annually. Professor Jameson, Mr. Weaver, and

others, have clearly proved that this majestic creature was coeval with man. A skull was discovered in Germany, associated with urns and stone hatchets; and in the county of Cork, a human body was exhumed from a wet and marshy soil, beneath a bed of peat eleven feet thick; the body was in good preservation, and enveloped in a deer skin, covered with hair, which there is every probability to conclude was that of the Elk. A rib of the Elk has also been found, in which there is a perforation, evidently formed by a pointed instrument while the animal was alive, for there has been an effusion of callus or new bony matter, which could only result from something remaining fixed in the wound for some time; such an effect, indeed, as would be produced by the head of an arrow, after the shaft was broken off.* There is, therefore, presumptive evidence that the race was extirpated by the hunter-tribes who first took possession of these islands.†

In the remarkable examples just cited, we have an interesting transition from the recent to the lost types of animal existence. 1st. Species extinct in these islands, but which are still living in other countries. 2dly. Animals whose absolute extinction is doubtful, but probable. 3dly. Species which have been entirely destroyed within the last few centuries. Lastly, Animals that were blotted out from the face of the earth by the early races of mankind.

* Jameson's Cuvier. † Appendix H.

7. Epoch of Terrestrial Mammalia.—We must now advance another step in the history of the past, and proceed from the consideration of what is known, to that which is unknown; and I shall restrict the subsequent divisions of this discourse, to the geological phenomena of the period immediately antecedent to the present; a period in which the earth appears to have teemed with enormous mammalia, and with which but few of the existing races were associated. Thus while the present may be termed the Modern or Human Epoch, that which forms the immediate subject of our investigation may be designated the Epoch of gigantic Mammalia.

8. Character of the Ancient Alluvial Deposites.—" When the traveller," says Cuvier, " passes over those fertile plains, where the peaceful waters preserve, by their regular courses, an abundant vegetation, and the soil of which is crowded by an extensive population, and enriched by flourishing cities, which are never disturbed but by the ravages of war, or the oppression of despotism, he is not inclined to believe that nature has also had her intestine wars, and that the surface of the globe has been overthrown by various revolutions and catastrophes. But his opinions change as he penetrates into that soil at present so peaceful; or as he ascends the hills which bound the plains. His ideas expand, as it were, with the prospect, and so soon as he ascends the more elevated chains, or follows the beds of

those torrents which descend from their summits, he begins to comprehend the extent and grandeur of those physical events of ages long past. Or if he examines the quarries on the sides of the hills, or the cliffs which form the boundaries of the ocean, he there sees, in the displacement and contortion of the strata, and in the layers of water-worn materials, teeming with the remains of animals and plants, proofs that these tranquil plains, these smooth un-broken downs, have once been at the bottom of the deep, and have been lifted up from the bosom of the waters; and every where he will find evidence that the sea and the land have continually changed their place."

In almost every part of the world, beneath the modern alluvial detritus, the nature and character of which were described in the former lecture, ex-tensive superficial beds of gravel, clay, and loam, are found spread over the plains, or on the flanks of the mountain chains, or on the crests of ranges of low elevation ; and in these accumulations of water-worn materials, are found immense quantities of the bones of large mammalia.* These remains belong principally to enormous animals related to the elephant, as the mammoth, mastodon, &c., and the hippopotamus, rhinoceros, horse, ox, deer, and many extinct genera and species ; while in caverns and

* The term *diluvium* is commonly applied to these ancient alluvial beds ; they are the newer *pliocene* in the classification of Mr. Lyell, as we shall hereafter explain.

fissures of rocks filled with calcareous breccia, the skeletons of tigers, boars, hyenas, and other carnivorous tribes, are imbedded. Remains of this kind exist, in such abundance, all over Europe, Asia, and America, that it is impossible to enumerate the localities; they are found alike in the tropical plains of India, and in the frozen regions of Siberia; while there is no considerable district of Great Britain in which some traces of these fossil bones do not occur.

9. CLASSIFICATION OF THE ORGANIC REMAINS. —Dr. Buckland, in his Bridgewater Essay, considers these remains as referrible to four divisions.

First. Land animals, drifted into estuaries or seas, and associated with marine shells, such as the Subappennine formations; the beds of gravel, sand, &c. provincially termed *Crag*, in Norfolk and Suffolk; loam and chalk conglomerate of Brighton cliffs; clay off Harwich and Herne Bay, and on the coast of Western Sussex.

Secondly. Terrestrial quadrupeds, imbedded with fresh-water shells; these strata have been formed during the same epoch as the above, at the bottom of fresh-water lakes; such are the lacustrine marls of the Val d'Arno.

Thirdly. Similar remains found in superficial detritus, spread over the surface of rocks of all ages. In beds of gravel near London; Petteridge Common, Surrey; and near Eastbourn, Sussex.

Fourthly. Osseous remains of carnivorous and

I

herbivorous animals in caverns and fissures of rocks
which formed part of the dry land, during the later
period of the same epoch. The caverns of Gaylenreuth, Kirkdale, &c. are examples.

Lastly. The relics in the osseous breccia, found
in the fissures of limestone on the shores of the
Mediterranean, in the Ionian Isles, in the rock of
Gibraltar, at Plymouth, and in the Mendip hills.*

Before I direct your attention to the fossils collected from those alluvial deposites, which are on
the table, and which comprise specimens from almost
every part of the world, it will be necessary to
review the leading principles of that science which
explains the structure of animal existence. Thus
while in our preceding investigations we referred to
Astronomy to dissipate the obscurity which shrouded
the earliest history of our planet, we are now led to
that most important department of natural knowledge, Comparative Anatomy, to enable us to restore the lost forms of animal existence. I shall
therefore point out to you the mode of induction,
employed by the scientific observer, in his investigation of the fossil remains of animals, and by
which he is enabled to ascertain the structure and
habits of those creatures which have long since
disappeared from the face of the earth.

10. COMPARATIVE ANATOMY. — To a person
uninstructed in this science, the specimens before

* Dr. Buckland's Bridgwater Essay, p. 94.

us would appear a confused medley of bones and
osseous fragments, impacted in the solid stone; and
the only knowledge he could derive from their examination would be the fact, that the stone was once
in the state of sand or mud, in which, while soft,
the bones became imbedded. But in vain would
he seek for farther information from these precious
historical monuments of Nature; to him they would
appear as unintelligible as were the hieroglyphics
of Egypt, before Young and Champollion explained their mysterious import. It is only by an
acquaintance with the structure of the living forms
around us, and by acquiring an intimate knowledge of their osseous frame-work or skeletons, that
we can hope to decipher the handwriting on the
rock, obtain a clue that will guide us through
the labyrinth of fossil anatomy, and conduct to
those interesting results, which the genius of the
immortal Cuvier first taught us how to acquire.
And here it will be necessary to enter upon the
consideration of those beautiful principles of the
relation of structure in organized beings, which
were first announced by that illustrious philosopher.

11. ADAPTATION OF STRUCTURE IN ANIMALS.
—The organs of every animal are parts of a machine,
all mutually dependent and admirably adapted for
the functions they are destined to perform; and
such is the relation of the several parts of the
machine with each other, that any variation in

one part, is constantly accompanied by some corresponding modification in another.   This mutual
adaptation of the several parts of the animal fabric
is a law of organic structure, which, like every
other induction of physical truth, has only been
established by patient and laborious investigation.
It is by the knowledge of this law that we are
enabled to re-assemble, as it were, the scattered
organs of the beings of a former state of the globe,
—to determine their place in the scale of animated
nature,—and to reason on their structure, habits,
and economy, with as much clearness and certainty,
as if they were still living and before us.   I will
demonstrate this proposition by a few examples.
Of all the solid parts of the frame the most obviously
mechanical are the jaws and teeth ; and as we know
in each instance the operations they are intended to
perform, this part of the animal structure affords
the most simple yet striking illustration of the principles I have just enunciated.

Tab. 10.—Skull of the Bengal Tiger.

12. Osteological Character of the Carnivora. — Let us examine the jaws of the skull

before us, that of a Bengal tiger.   We perceive
that there are cutting teeth in front,—sharp fangs
on the sides,—and molar, or bruising, or crushing
teeth, in the back part.   The molar rise into sharp
lanciform points, and over-lap each other in the
upper and lower jaw, like the edges of a pair of
shears; and the teeth are externally covered with
a thick crust of enamel.   This is evidently an
apparatus for tearing and cutting flesh, or for
cracking bones; but is not suited for grinding the
stalks or seeds of vegetables.   The jaws fit together
by a transverse process, which moves in a corresponding depression in the skull, like a hinge. (Tab.
10. 3.)   They open and shut like shears, but admit
of no grinding motion; this, then, is such an articulation as is adapted for a carnivorous animal;
and every part of this instrument is admirably
fitted for its office.   But all these nice adjust-
ments would be lost, were there not levers and
muscles to work the jaws,—were not each part of
the animal frame adapted to all the other parts,—
and were not the instincts and appetites of the
animal such as are calculated to give to this apparatus its appropriate movements.   Let us reverse
the order of our argument,—let us assume that the
stomach of an animal be so organized as to be
fitted for the digestion of flesh only, and that flesh
recent,—we should find that its jaws would be so
constructed as to fit them for devouring live prey,
—the claws for seizing and tearing it,—the teeth

for cutting and dividing it,—the whole system of its
powers of motion for pursuing and overtaking it,—
the organs of sense for discovering it at a distance,—and the brain endowed with the instinct
necessary for teaching the animal how to conceal
itself, and lay snares for its victims. Such are the
general relations of the structure of carnivorous
animals, and which every being of this class must
indispensably combine in its constitution, or its race
cannot exist. But subordinate to these principles,
are others connected with the nature and habits
of the prey upon which the animal is intended to
subsist, and thence result modifications of details
in the forms which arise from the general conditions. Thus, in order that the animal may have
the power requisite to carry off its prey, there must
be a certain degree of vigour in the muscles which
elevate the head; and thence results a determinate
form in the vertebræ or bones from which these
muscles originate; and in the back of the head in
which they are inserted. That the paws may be
able to seize their prey, there must be a certain
degree of mobility in the toes, and of strength in
the claws, and a corresponding form in all the
bones and muscles of the foot. It is unnecessary to
extend these remarks, for it will easily be seen that
similar conclusions may be drawn with regard to
all the other parts of the animal. In the tiger we
have a familiar illustration of what has been advanced.

13. STRUCTURE OF THE HERBIVORA. — In animals which are destined to live on vegetables we have the same mutual relations; the sharp fangs of the teeth are wanting, the enamel is not all placed on the top of the teeth as in the carnivora, but is arranged in deep vertical layers, alternating with bony matter; and this arrangement, in all states of the teeth, secures a rough grinding surface, as in the horse and the elephant.   This fossil tooth of a

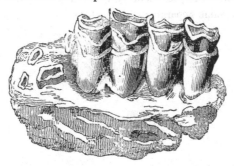

TAB. 11.—TOOTH OF A RUMINANT IN OSSEOUS BRECCIA.

ruminant from Cerigo will serve as an illustration. The flat molar teeth are not formed for cutting, but for mastication, and the jaws are loosely articulated together so as to allow of a grinding movement: had the socket and corresponding part of the jaw been the same as in the tiger, the tooth could not have performed its office.   Again, I might proceed in the argument, and show the adaptation

of the muscles of the head to the apparatus here
described; and beginning with the jaw review the
whole animal frame, and demonstrate how all its
parts are alike wonderfully constructed and fitted
together, to perform the functions necessary for the
being to whom it belongs.

14. STRUCTURE OF THE RODENTIA, OR GNAW-
ERS.—If we now examine the jaw of some intermediate order, we shall perceive new adaptations
of the same apparatus.

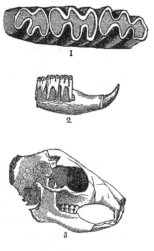

TAB. 12.—SKULL AND TEETH OF RODENTIA.

FIG. 1. Molar Teeth of the Upper Jaw of a Florida Rat (*Arvicola
Floridana*) magnified; seen obliquely. FIG. 2. Left side of the Lower
Jaw, of the natural size. FIG. 3. Skull of the Squirrel.

15. SKULL AND TEETH OF RODENTIA.—Thus the
animals called rodentia or gnawers, have long sharp
cutting teeth, like pincers; hence the rat can very
speedily gnaw a hole through a board, and the squirrel in a nut, in consequence of the exquisite adaptation of their teeth for these operations. In this
skull of the squirrel (Tab. 12, Fig. 3) you may perceive that the front teeth are of enormous size, as
compared with the molar, and that they lock together in such a manner as to render a grinding
movement impossible; a new adjustment has, therefore, been supplied,—the lower jaw is so adapted as
to work in the skull neither in a transverse nor in a
rotatory direction, but lengthwise, like the action
of a carpenter using his plane. And if you watch
a rabbit while eating, you will perceive that its teeth
move backward and forward precisely as I have described. The enamel of the molar teeth (see Tab. 12,
Fig. 1) is placed vertically and transverse to the
jaw, so as to form an admirable grinding surface.
But this is not the only variation of structure observable in the teeth of these animals. The incisors
being implements of continual use, are renewed
by perpetual growth, and there is a special provision for their support in a bent socket. The
enamel is unequally distributed round the tooth,
being very thin behind and thick in front, by which
means the cutting edges are always preserved. By
the very act of gnawing, the hinder part of the
incisor wears away quicker than the fore part, and

thus it is that a sharp inclined edge is maintained, like that of an adze or chisel, and which is the very form required in the economy of the animal. The skull of- the common rabbit or hare will exemplify these remarks. These are but a few of those admirable adaptations of means to ends which are observable throughout the various classes of organized beings : but the brief space of a single lecture will not allow me to be more diffuse, and I trust it is unnecessary to offer further remarks, to show that the conclusions of geologists, as to the ancient inhabitants of our globe, are not vague assumptions, as those unacquainted with the science they attempt to impugn assert, but the legitimate deductions of laborious and patient investigation. A few teeth and bones—sometimes but a single relic of this kind — are the elements by which the comparative anatomist is enabled, not only to restore the forms of creatures now banished from the face of the earth, but also to ascertain their habits and economy, and even arrive at positive conclusions respecting the nature of the country of which they were once the inhabitants. For if we find the remains of animals which lived on vegetables, it follows that there must have been vegetables for their subsistence, and a condition of nature calculated for the growth of vegetable productions; a soil fitted for their existence, and a country diversified by hills, and valleys, and plains, with streams and rivers to carry off its superfluous waters. The same laws, under certain modifications, apply to other classes of beings. Thus in birds, the form of the feet is modified according to the habits of the different orders. In the parrot, the claws are adapted to climb trees and perch on the branches; but in the eagle they are widely different, for its talons are constructed to lacerate and tear its prey. The feet of aquatic birds are formed like a paddle or oar, to enable them to make their way through the water; those of birds that frequent marshes, have a great expansion, like a tripod, that they may move over the unstable surface of the morass; while in species destined to inhabit sandy deserts, as the ostrich, the feet present a corresponding change.

Thus we find that every vertebrated animal has a solid and durable skeleton, or osseous support, formed upon one general plan, but modified in almost endless variety, in the relative magnitude, situation, and aspect of the different parts, so as to adapt itself to the various habits and functions of the diversified forms of animal life. In short, that the Author of nature has by these changes varied the same general fabric in innumerable ways; bestowed upon it a thousand different instincts and passions; adapted it to every element and climate; and to every possible variety of food and mode of existence.

From a knowledge of these principles of the co-relation of the different parts of every organized being, which I have thus attempted to elucidate, we may understand how the scientific observer can reconstruct the entire animal fabric : and we are now prepared to enter upon that department of Geology called Palæontology, or the science which relates to the fossil remains of the beings which were the inhabitants of our planet in former ages.

16. FOSSIL BONES.—As the bones are the least perishable parts of the animal structure, they become the most frequent, and often the only indications of the zoological character of the more ancient epochs. Occasionally very delicate parts, such as the tunic of the eye, the membranes of the stomach, and the wings of insects, are preserved in a fossil state, examples of which we shall hereafter adduce. In the more ancient deposites, the bones are generally mineralized, and no longer possess the white and glossy appearance of the recent skeleton ; but those which occur in the superficial gravel, and in caverns, are commonly of a porous and earthy character, like bones that have lost a portion of their animal matter by being buried in a dry and loose soil.

The animals whose fossil remains I now proceed to describe, may be separated into two classes— the HERBIVORA, whose bones occur in the gravel and marl,—and the CARNIVORA, which are found in fissures and caverns.

17. FOSSIL ELEPHANTS, OR MAMMOTHS.*—

* From the Arabic *behemoth*, signifying *elephant.*

I will first notice the fossil remains of the animals of the elephantine family. These occur in great abundance, and are very generally distributed. In the earlier ages, their colossal bones were supposed to belong to gigantic races of mankind, and hence the tradition of giants possessed by every country in Europe : nor need we smile at the ignorance and credulity of our ancestors, for, not many years since, a fossil tooth of an elephant, which was discovered in digging a well in this town, was supposed to be a petrified cauliflower !—In Russia, particularly in Siberia, the bones of fossil elephants are found throughout all the low lands, and in the sandy plains, (but not in the elevated primary chain of hills,) stretching from the borders of Europe to the nearest extreme point of America, and south and north from the base of the mountains of central Asia, to the shores of the Arctic sea. Within this space, which is scarcely inferior in extent to the whole of Europe, fossil ivory is every where to be found ; and the tusks are so numerous, and so well preserved, especially in Northern Russia, that thousands are annually collected, and form a lucrative article of commerce, being exported as ivory for turning. In Siberia alone, the remains of a greater number of elephants have been discovered, than are supposed to exist at the present time all over the world. In a low island in the Frozen Sea (72° north latitude) bones of mammoths are seen imbedded ; and they also abound in an iceberg on the north-west angle of the American continent, close to Behring's Straits.

18. MAMMOTH AND RHINOCEROS IMBEDDED IN ICE.—But the most remarkable fact relating to these remains, is the preservation, not merely of the bones, but of the flesh and skin, in short, of entire animals, in ice-bergs and in frozen gravel! In 1774, near Vilhoui, the carcase of a rhinoceros was taken from the frozen sand, where it must have been concealed for ages, the soil of that region being always frozen to within a few inches of the surface. The carcase was a complete natural mummy, part of the skin being still covered with long hairs, and forming a warmer covering than that of the African rhinoceros. The discovery of a mammoth, under similar circumstances, is still more interesting. It appears, that towards the close of the last century, a Tungusian fisherman observed in a cliff of ice and gravel, on the banks of the river Lena, a shapeless mass, the nature of which he was unable to determine. In the course of the next year it was more visible, and on the third a large tusk was seen projecting from the ice-cliff, and at length became detached. On the fifth year, an early thaw set in, and the entire carcase of a mammoth was exposed, and at length fell upon the ground. It was nine feet high, and about sixteen feet in length ; the tusks were nine feet long. The flesh was in such a state of preservation, that it was devoured as it lay by wolves and bears, and the

hunters fed their dogs with the remains. The skin was covered with hair, consisting of black bristles, thicker than horsehair, and fifteen inches in length; wool of a reddish brown, and hair of a fawn colour; and with a mane on the neck. Upwards of 30 lbs. of hair were collected, and specimens of it are preserved in the Hunterian Museum. The ear remained dry and shrivelled; the brain and even the capsule of the eye were preserved! the bones and part of the integuments, and a considerable quantity of the hair, are in the Museum of Natural History at St. Petersburgh. The accompanying sketch represents the skeleton in its present state.

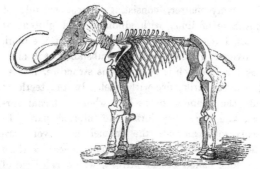

TAB. 13.—MAMMOTH IMBEDDED IN FROZEN GRAVEL IN SIBERIA.

The occurrence of large mammalia, in latitudes where but few forms of animal life can now possibly find the means of subsistence, is a fact of so much interest, that I must indulge in a few additional remarks. You are aware that the existing elephants belong to two species, the African, which occurs as far south as the Cape of Good Hope; and the Asiatic, which is limited to 31° north latitude. They are distinguished by certain characters; but those which more especially relate to our present inquiry, are the peculiarities of the teeth.

19. TEETH OF RECENT AND FOSSIL ELE-PHANTS.—The teeth of animals are formed of three distinct substances, which are variously disposed in different orders, according to the habits and economy of the species; a fact to which I alluded when treating of the distinguishing character of the rodentia, &c. The nucleus of the tooth is formed of a bony matter, consisting almost entirely of phosphate of lime, with albumen, and gluten; it is called ivory. This central portion of the tooth is covered by the enamel, a substance still more dense, and which is of a fibrous structure, and so hard as to strike fire with steel. In the teeth of man, the enamel covers the whole external surface, and the ivory forms the internal part. In herbivorous animals the enamel and ivory are intermixed; and there is in some genera, a third substance called *crusta petrosa*, which is a kind of yellowish, opaque ivory. These three substances enter into the composition of the teeth of the elephant, and in the masticating surfaces their intermixture is apparent; they are differently disposed in the two species. In the African elephant (Tab. 14, fig. 2,) the worn surface of the molar

teeth presents a series of lozenge-shaped lines of enamel, having the ivory on the inner margin of the ridges, and being surrounded by the *crusta petrosa*.

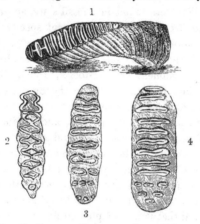

TAB. 14.—TEETH OF RECENT AND FOSSIL ELEPHANTS.

Fig. 1. Fossil Tooth of an Elephant from Brighton Cliff. Fig. 2. Crown of a Tooth of the African Elephant. Fig. 3. Grinding surface of a Tooth of the Mammoth. Fig. 4. Grinding surface of the Asiatic Elephant.

In the Asiatic species (fig. 4) the enamel forms narrow transverse bands; and the tooth of the mammoth, or fossil elephant, (Figs. 1 and 3,) has an analogous, but somewhat different distribution. It is obvious that the structure here exhibited, is fitted for the grinding of vegetables; for the three substances, being of different degrees of hardness, produce by their unequal wearing, a constant rough

K

surface for trituration.* The elephant has but four
teeth in each jaw; the deficiency of prehensile
teeth being supplied by that wonderful organ, the
trunk. The teeth found in a fossil state, appear to
be distinct from either of the recent species; but
they are more nearly related to the Indian or Asiatic, than to the African species, as you may observe
by these specimens from Siberia, India, North
America, and from the cliffs on the Sussex coast.
In some examples the teeth are water-worn, but
most commonly are very perfect, and exhibit no
marks of attrition. From a careful review of all
the characters of the fossil elephant, or mammoth
of Siberia, Cuvier determined that the species was
now extinct; that the structure of the teeth, configuration of the skull, and its hairy and woolly
skin, proved that it was adapted to live in a colder
climate than that in which the Asiatic species could
exist; and he infers that the animals originally inhabited the countries where their remains are now
found imbedded; and that the preservation of the
carcases in ice, proves that the change in the temperature of the climate was sudden, and has since
remained unaltered. Mr. Lyell offers an ingenious
solution of this curious problem. He supposes that
a large region of central Asia, perhaps the southern
half of Siberia, may have enjoyed a climate mild
enough to have admitted of the existence of the

* See Dr. Roget's Bridgwater Treatise, for a lucid and most
interesting Essay on the teeth of animals.

extinct elephants; for vegetation may be found in lat. 40° and 50° north: and he infers from the physical geography of the country, that the whole tract from the mountains to the sea may have been upraised like Sweden ; and that the refrigeration of the north-east of Asia, and its present physical condition, have been the result.

Time will not permit me to dwell at length on other discoveries of fossil elephants, but I will briefly mention a few instances in our own country. On the coasts of Norfolk and Suffolk, so many teeth of fossil elephants have been collected, that Mr. Woodward (the author of " The Geology of Norfolk") has calculated that they must have belonged to above 500 individuals. At Walton, in Essex, and at Herne Bay, bones and tusks have been found. But by far the most extraordinary collection of the remains of British fossil elephants that I ever beheld, is in the possession of Mr. Gibson, of Bow, near London ; it contains skulls, tusks, and teeth, from the sucking animal to the adult, and in a marvellous state of preservation ; the whole of which were discovered in Essex. In the highly interesting museum of W. D. Saull, Esq. of London, many fine elephantine remains are also preserved. On the western coast of Sussex, and in the neighbourhood of Arundel, and at Patcham, and in this town, teeth and bones of elephants have at different times been exhumed. At Brighton the teeth are found in a deposit of water-worn materials, consisting of loam, chalk, and broken flints, resting on a bed of shingle which covers the chalk.* In the conglomerate, of which I have already spoken, (Tab. 7. p. 90,) as well as in the superincumbent deposit, the teeth of elephants, with bones and teeth of a species of deer, and horse, and bones of Cetacea, occur, and are associated with marine shells. When these remains were imbedded, this part of the English coast, as well as the opposite shores of France, must have formed part of the boundary of a bay or estuary, of a country inhabited by large Mammalia; for similar fossils are found in a deposit of a like character, along the French coast.

20. THE MASTODON.—In various parts of North America, there are marshy tracts abounding in salt or brackish waters, which are frequented by deer, and other animals ; a circumstance from which they have acquired the American name of *Lick*. In these morasses vast quantities of bones of gigantic terrestrial Mammalia have been discovered. The spot most celebrated for these remains in Kentucky, is called Big-bone Lick, and is situated to the south-east of the Ohio, in the midst of a group of low hills, and is traversed by a small stream of brackish water. The bottom consists of a black fetid mud, intermingled with sand, and traces of vegetable matter. In this bog, bones of great magnitude occur in profusion. Some of them are

* See Geology of the South-East of England, p. 32.

referrible to the fossil elephant, but others, as you
may observe from the specimen before us, must
have belonged to a creature not less gigantic, but
with very different characters.

TAB. 15.—TOOTH OF THE GREAT MASTODON.
(*From Professor Silliman of Yale College.*)

These teeth, you will in an instant observe, are
very dissimilar to those of the elephant; they are
composed of ivory and enamel only, and the enamel,
which is very thick, is spread over the crown of
the tooth, which, when unworn, is divided into
several transverse tubercles, or processes, each of
which is subdivided into two obtuse points; from
this character of the teeth the name of MASTODON
(from two Greek words, signifying mammillary
teeth,) has been given to the creature from whom
they are derived. These teeth have no relation to
those of the carnivora; for although they have an

external investment of enamel like those of the
tiger, yet they are destitute of the longitudinal,
serrated, cutting edge ; and in those which are worn,
the protuberances become truncated into a lozenge
form. The structure is similar to that of the hog
and the hippopotamus, and is fitted for the bruising
and masticating of crude vegetables, roots, and
aquatic plants. The remains of the Mastodon
have been found throughout the plains of North
America, from north of Lake Erie to as far south
as Charleston, in South Carolina; they have been
also discovered on the Continent, and in the Crag
of Norfolk, in England. Here are examples from
the banks of the Ohio, of the Hudson, and from
Big-bone Lick, presented to me by Professor Silliman ; this is an example of a young perfect tooth,
and this of a very old animal, for the surface is
almost worn flat by use. The remains of the
Mastodon are found at moderate depths, with no
marks of detrition, and therefore the animals must
have lived and died in the country where their
relics are entombed. The skeletons of the great
Mastodon found in bogs in Louisiana are in a
vertical position, as if they had sunk in the mire;
and one discovered in New Jersey, forty miles to
the south of New York, was found in black earth,
in the same position, the head being on a level with
the surface of the soil. There is an entire skeleton
of the Mastodon in the museum of Mr. Peale, in
Philadelphia: it is fifteen feet long and eleven feet

high; and by this specimen it has been ascertained
that the Great Mastodon, or animal of the Ohio, as
it has been called, was not unlike the elephant, but
somewhat longer and thicker. It had a trunk or
proboscis, tusks, and four molar teeth in each jaw,
and no incisors. From the nature of its food, as
shown by the structure of the teeth, it must have
frequented marshy tracks, but it was undoubtedly a
terrestrial animal. In the midst of a collection of
these bones imbedded in mud, was found a mass of
little branches, grass, and leaves, in a half bruised
state, and among these was a species of reed, common in Virginia; the whole appeared to have been
enveloped in a sack, probably the stomach of the
animal. In another instance, traces of the trunk or
proboscis were observed. The tusks are composed
of ivory, and vary in their curvature like those of
the elephant. The bones of this colossal quadruped
are found remarkably fresh and well preserved;
but they are generally impregnated with iron, and
have evidently been buried in the earth for ages.
No living instance of this creature is on record, and
no doubt can exist that its race is totally extinct.
The Indians believe that men of similar proportions
were coeval with the Mastodon, and that the Great
Spirit destroyed both with his thunder.* There
are several species of the Mastodon, some of which
have been found in North America only, and

* Cuvier. See an admirable English Epitome of Cuvier's
Fossil Animals, by Edward Pidgeon, Esq. 1 vol. 8vo. 1833.

others in Europe. That eminent philosopher Baron Humboldt discovered a tooth of the Mastodon near the volcano of Imbaburra, at an elevation of 1,200 fathoms. The turquoises of Simone are composed of mammoth bones, impregnated with some metallic oxide. In conclusion I will mention, that a very fine skull, with teeth, of the Great Mastodon, from Big-bone Lick, has lately been placed in the British Museum, and is well worthy your notice when visiting that now magnificent collection. This specimen, which was purchased for 150 guineas, consists of the cranium with two perfect grinders, and the sockets of the other two. The length of the skull, from the occiput to the socket for the tusks, is 36 inches.

21. MASTODONS FOUND IN THE BURMESE EMPIRE.—I now request your attention to the remains of a species of Mastodon which, from the structure of the teeth, fills up, as it were, the interval that separates the Mastodon from the Elephant, and which has been named by Mr. Clift, the *Mastodon elephantoides*. The teeth, which, together with the collection of fossil bones and wood before us, I owe to the liberality of Mr. Craufurd, present characters very peculiar; for while their structure is similar to that of the Great Mastodon, the form of the ridges in which the crown of the teeth is disposed, resembles those of the elephant; and the grinding surface of the tooth, when worn, would bear an analogy with that of the African elephant.

These teeth, together with bones and teeth of the
hippopotamus, rhinoceros, horse, tapir, ox, antelope,
hog, gavial, fresh-water turtle, &c. and silicified
wood, are part of an extensive collection formed about
ten years since by Mr. Craufurd, on his mission to
Ava. Descending the river Irawaddi, his steam-
boat, owing to the shallowness of the water, ran
aground, between Prome and Ava, about 20° north
latitude, near some Petroleum wells, where the bank
of the river presents a cliff 80 feet high; and on
the strand were observed masses of petrified wood,
and vast quantities of bones. The adjacent country
is formed of low, sterile sand-hills, intersected by
ravines, with beds of gravel, which are here and
there cemented into a breccia by iron and carbonate
of lime, by the process which was explained in the
former lecture. Scattered over the surface, and in
some instances lying loose in the sand, in others
half buried, were masses of silicified wood, and
fragments of bones, which had become exposed,
by the removal of the sand by winds or rains.
The bones, as you may perceive, in these incrusted
specimens, were more or less invested with a hard
calcareous, siliceous conglomerate, which appears
to be a mere local concretion, from the consolidation of the loose sand by ferruginous and
calcareous infiltrations. The natives who assisted
Mr. Craufurd's party in collecting these remains,
believed that they were the bones of giants who
had warred against Vishnu, by whom they had

been destroyed.   On these important discoveries of
Mr. Craufurd, Dr. Buckland* calls attention to the
remarkable fact, that in the twelve chests full of
osseous remains, not a fragment belongs to the
elephant, tiger, and hyena, which abound in India;
while of the extinct Mastodon, evidence is afforded
that it must once have swarmed in the districts
bordering on the Irawaddi.

22. THE SIVATHERIUM.† — The flanks of a
range of hills belonging to the Sub-Himalaya
Mountains, between the river Sutlej and the Ganges,
are covered by beds of concretionary sandstone,
conglomerate, and loam, bearing a close analogy to
those of Ava we have just examined.   These hills,
which are called the Sivalik, (from Siva, an Indian
deity,) rise to an altitude of from one to three
thousand feet above the level of the sea.   In these
deposites occur immense quantities of the fossil
teeth, and bones of the elephant, mastodon, hippopotamus, rhinoceros, elk, ox, horse, deer; and of
several carnivorous animals, crocodiles, gavials, and
fresh-water turtles; with fluviatile shells, and remains of fishes.   Extinct species of *monkey* and
camel have also been found.   These interesting discoveries were made by Captain Proby Cautley, of
the Bengal artillery, and Dr. Falconer, who, with
an energy and perseverance beyond all praise, have

* See an interesting Memoir on the Bones from Ava, by
Dr. Buckland.   Geol. Trans. vol. ii.   New Series.

† From *Seva,* an Indian deity, and *therium,* wild animal.

followed out their researches, and transmitted magnificent collections of these remains to England. The valuable specimens before us were, with great liberality, sent to me by Captain Cautley, to whom I am an utter stranger; among them you may observe the same species of mastodon as that of Ava; with bones of the horse, rhinoceros, hippopotamus, gavial, and a fine skull of the fossil elephant, with all the four teeth perfect. But the labours of these naturalists have been yet more richly rewarded by the discovery of the skull, and other parts of the skeleton, of a creature hitherto unknown; one that forms, as it were, a link between the ruminants and the large pachydermata. From the skull, which is remarkably well preserved, it is ascertained that the animal had four horns, and was furnished with a proboscis; that it was larger than a rhinoceros, and combined the horns of a ruminant, with the characters of the pachydermata. The discoverers have named it the Sivatherium. This animal, when living, must have resembled an immense antelope, or Gnu; with a short and thick head, an elevated cranium, crested with two pairs of horns; the front pair were small, and the hinder large, and set quite behind, as in the aurochs. With the face and figure of the rhinoceros, it must have had small lateral eyes, great lips, and a nasal proboscis; these inferences have been deduced from certain anatomical characters exhibited by the fossil bones, but upon which I must not now enlarge.

23. THE MEGATHERIUM. (*Megas*, great, and
Therium, wild beast.)—The Pampas, those immense
plains of South America, on the south bank of the
river Saladillo, which present a sea of waving grass
for 900 miles, are principally composed of alluvial
loam and sand, containing fresh-water with marine
shells, and were once, like Lewes Levels, a gulf, or
arm of the sea. In these alluvial deposites, enormous
bones have been frequently discovered. Towards the
close of the last century, an almost entire skeleton of
a gigantic animal was dug up, at the depth of 100
feet, in a bed of clay, on the banks of the river
Luxor, about four leagues W.S.W. of Buenos Ayres.
This skeleton was sent to the museum at Madrid,
in 1789, where it now remains. It is described and
figured by Cuvier, under the name of the Megatherium. In 1832, my friend, Sir Woodbine Parish,
with considerable labour and expense, collected
many parts of a skeleton of a similar creature from
the Salado, actually diverting for a time the river
from its course, that he might disinter these precious
relics, which he has since deposited in the Hunterian
Museum of the Royal College of Surgeons. But,
before I enter upon a description of these fossils, it
will be requisite to notice the remains of an animal
of analogous structure, which has been discovered
in the saltpetre caves in Virginia and Kentucky, and
which, from the size of the unguical or claw-bones,
has been named the Megalonyx.

24. The Megalonyx.*—I have placed upon the table, models of all the bones which are now preserved in the museum of Philadelphia, and for which I am indebted to an eminent physician and geologist of that city, Dr. S. G. Morton, the author of the most valuable treatise that has appeared on the Fossils of the United States.† The late American President, Jefferson, who first described the remains of this animal, inferred, from the form and magnitude of the claw-bone, that the creature was a carnivorous animal of colossal proportions. But Cuvier, by his profound knowledge of the principles of anatomy, determined, from certain characters of the articulating surfaces,‡ that the animal was related to the *Bradypus*, or sloth. I will endeavour briefly to explain to you the mode by which this induction was obtained. The paws or feet, both of the canine and feline tribes, are armed with claws; in the former, the nails are thick and coarse as in the dog, wolf, &c. and fitted to bear the friction and pressure incident to a long chase; while in the cat tribe, on the contrary, they are curved and sharp, which qualities are preserved by a peculiar mechanism. The last bone which supports the claw is placed

---

* *Megas*, great; *onyx*, claw. See a "Description of the Fossil Bones of the Megalonyx," in Dr. Harlan's Medical and Physical Researches.

† *Synopsis of the Fossils of the Cretaceous Group.*

‡ That surface of the bone which forms a joint with another corresponding bone.—G. F. R.

laterally to the penultimate bone, and is so joined
to it that an elastic ligament draws it back, and
raises the sharp extremity of the claw upwards;
and the nearer extremity of the farthest bone
presses the ground in the ordinary running of the
animal, while the claw is retracted into a sheath:
but when the creature makes a spring and strikes,
the claws are uncased by the action of the flexors
or bending tendons. In the Bengal tiger, the claws
are so sharp and strong, and the arms so powerful,
that they have been known to fracture the skull of
a man, by a single touch in the act of leaping over
him.* A cat affords a familiar illustration of this
peculiarity of structure ; when pleased, its claws are
retracted, and when angry they are thrown out.
In the claw of the Megalonyx there is no such lateral
provision, which is necessary for its retraction, and
the point could not have been raised vertically, as
in the cat, so as to have permitted it to touch the
ground without injury. The articulating surface is
double, that is, there is a ridge or spine in the middle, and it must, therefore, have moved like a hinge.

25. THE SLOTH.—There is among recent animals an order called *Tardigrada*, from their feeble
power of progression—these are the *Paresseux* or
Sloths ; which have large toes, and long nails, of a
construction similar to those of the fossil. Their
nails are folded up, but in a very different manner
from those of the cat; they just enable the animal

* Sir C. Bell.

to walk, in the same way as if our fingers were folded under the palms of the hands. This is a specimen of the *Bradypus* (slow-footed) *tridactylus* (three-toed) from South America, and which is called the *Ai*, from its peculiar cry. The arms, you perceive, are double the length of the legs, and, from the construction of the limbs, the animal, when it walks, or rather crawls on the ground, is obliged to drag itself along on its elbows. But these crea‑ tures are destined to inhabit trees; their proper element is on the branches, and they can pass from bough to bough, and from tree to tree, with a rapidity which soon enables them to lose themselves in the depths of the forests. They live on the leaves and the young shoots, and, unless disturbed, never quit a tree till they have stripped off every leaf. To avoid the trouble of a descent, they drop to the ground, previously coiling themselves into a round ball, in which state, while attached to the branch, they may be taken alive. This brief notice of the habits and economy of the Sloth, points out the necessity for a peculiarity of structure in its nails. The monkey leaps and swings himself from tree to tree, and catches at will the branches or the trunk. But the sloths do not grasp; their claws are mere hooks to hang by, and their great strength is in their arms. They never unfix one set of hooks until they have caught a secure hold with the other, thus hanging by their arms and legs, while their bodies are pendant; and they sleep in the same position.

In the bones of the arm of the *Megalonyx*, we find
a corresponding analogy with those of the Sloths.
The humerus, or arm-bone, has a long internal
condyle for the origin of very large muscles to move
the enormous claws; and there is a foramen or
opening for the passage of nerves and blood-vessels,
to protect them from the pressure to which they
would be exposed from the powerful muscular action; while the radius is so constructed as to allow
of a rotatory motion of the arm. With the bones
of this animal were found masses of osseous polygonal scales, like mosaic work; and it is supposed that
the original was covered with an armour resembling
that of the armadillo.

I now proceed to the consideration of the Megatherium. This creature was about seven feet high,
and nine long, and therefore larger than the largest
rhinoceros; but this comparison by no means conveys a proper idea of its bulk, since its proportions
are perfectly colossal, the thigh-bone being three
times as large as that of the elephant, and the pelvis
or haunch-bone, twice the breadth. It possessed no
incisor teeth, and the molars or grinders are seven
inches long, of a prismatic form, and, like those of
the elephant, composed of ivory, enamel, and *crusta
petrosa*, or cement. They are so formed, that the
crown of the teeth always presents two cutting,
wedge-shaped, salient angles. As in an adze, a plate
of steel is placed between two of iron, so as to pro-
ject in a line, in like manner these teeth have in the

centre, a cylinder of ivory, which is protected by a plate of enamel, and has an external coating of *crusta petrosa ;* these teeth are, therefore, admirably adapted for cutting and bruising vegetable matter. The entire fore-foot is about a yard in length, and the claws are set obliquely to the ground, like those of the mole; a position which would render them digging instruments of great power. The pelvis measures five feet in width, and the sacral aperture of the spinal marrow is one foot in circumference ! This enormous size was suitable to the habits of an animal requiring to maintain an upright posture for a considerable time, and to employ its fore-feet in digging. As Dr. Buckland has so admirably elucidated the structure and habits of this enormous being of the ancient world, and as his work is, or ought to be in every library, I will not dwell on other important peculiarities in its osteology, but content myself with stating, that the *Megalonyx* and *Megatherium* were intermediate between the sloths, armadillos, and ant-eaters. The Megatherium, with the head and shoulders of the sloth, combined in its legs and feet an admixture of the characters of the armadillo and ant-eater. Both the megalonyx and megatherium were herbivorous, but they were not capable of climbing, even had there been trees that could have supported their enormous weight: their food, like that of the armadillos, must have consisted of roots and stems of succulent vegetables, which the peculiar structure of their feet enabled them to dig up with

L

facility. Like their recent types, they are limited in their geographical distribution to nearly the same regions of the new world.*

26. Fossil Hippopotamus, Rhinoceros, Horse, &c.—With the fossil remains of the Mammoth, Elephant, and other large mammalia, the teeth and bones of the Hippopotamus, Horse, Elk, Ox, and Auroch, are very commonly associated. In the Vale of Arno, in Italy, immense quantities of the teeth and bones of Hippopotami are found. On the table before us are specimens from that locality; as well as molars and incisors of a young animal from Huntingdonshire, presented by W. D. Saull, Esq. F. G. S.; and tusks and bones, dug up a few months since at Southbourn, in Sussex. Bones of this animal also occur in alluvial deposites near Rome; and here are examples, collected by the Marquis of Northampton. Among the objects sent me by Captain Cautley, from India, are several fine portions of jaws, with teeth, belonging to a Hippopotamus (*H. Sivaliensis.*) Several extinct species of Hippopotamus have been determined by Baron Cuvier, one of which was not more than half the size of the common species. The bones and teeth of the Rhinoceros are constantly associated with those of the fossil Elephant; and in this country they occur in superficial gravel and loam: these examples of teeth were discovered in Surrey.

See Dr. Buckland's Bridgewater Essay.

But the most extraordinary and interesting fact, relating to the fossil Rhinoceros, is the discovery of the entire carcass, with the skin, in frozen sand, on the banks of the Wilaji, in Siberia. The head was extremely large, and sustained two very long horns; it had no incisors; the body was covered with brown hair, particularly on the limbs; and the general form of the animal was lower and more compact than the living species.

The teeth and bones of one or more species of Horse, occur very constantly with those of the larger and extinct Pachydermata; in these examples of the conglomerated shingle from Brighton cliffs, the coffin, pastern, and cannon bones, as they are termed, are imbedded; in some instances the cavities of the long bones are filled with crystallized carbonate of lime.

In addition to the animals we have already noticed, the deposites now under our examination contain many lost species of ruminants and other orders of Mammalia. The fossil remains of an animal resembling the musk-ox are found with elephant's bones in Siberia; an extinct species of fallow-deer in Scania; of roe-buck and reindeer in France; and gigantic oxen, aurochs,* deer, &c. in our own country. As Dr. Buckland has given

* Auroch, a recent species of wild bull or buffalo, distinct from the common ox. The horns of the fossil ox are sometimes of enormous size: Mr. Parkinson had a pair in which the length of the horn was 2 feet 7 inches.

a list of the different animals that have been discovered in these beds, I will content myself with
the remark, that while the fossil pachydermata,
such as the elephant, rhinoceros, &c. belong to
genera which inhabit torrid climes, the ruminants
are of genera which at the present time are natives
of northern latitudes.

TAB. 16.—THE DINOTHERIUM.

27. THE DINOTHERIUM.*—I shall conclude my
remarks on the large Mammalia by a description
of a gigantic creature of a very peculiar character,
whose bones occur with those of the Mastodon,
Elephant, and other animals which we have already
examined; as well as with the remains found in more
ancient deposites. In various parts of the south
of France large molar teeth, resembling in their
form and structure the teeth of Tapirs, have been

* _Dinos_, terrible ; _therium_, a wild beast.

occasionally found; and they are described by
Baron Cuvier under the name of the " Gigantic
Tapir." These are models of the principal specimens
deposited in the Museum at Paris, presented to me
by Baron Cuvier, together with others which I shall
place before you on a future occasion. Subsequent
discoveries in Bavaria, Austria, and particularly at
Eppelsheim, about twelve leagues south of Mayence,
have made us acquainted with the form and structure of the original, which appears to have been
one of the largest of lacustrine animals, the skeletons
showing that some individuals were eighteen feet in
length! The scapula, or shoulder-blade, was like
that of the mole, and the fore-leg must therefore
have been adapted for digging up the earth. The
most extraordinary deviation from ordinary types
consists, however, in the curved tusks, which are
fixed in the lower jaw in a downward direction, as
those of the Walrus are in the upper; the lower
jaw is four feet in length. It is certain that the
creature had a proboscis, because it possesses no
incisor teeth with which to seize its food, and the
jaws do not even close together in front; and from
the structure of the anterior portion of the cranium, and the disposition of the nasal fossæ; the
tusks were probably weapons of defence, like those
of the elephant. The above drawing, (Tab. 16,)
from a restoration by M. Kaup, an eminent German naturalist, represents the supposed form of
the original creature. It would appear that the

Dinotherium was nearly related to the Hippopotamus, forming a link between the Cetacea and Pachydermata, or large terrestrial Mammalia; and that it was an herbivorous aquatic animal, and inhabited lakes and marshes.

28. FOSSIL CARNIVORA IN CAVERNS. — We have passed in review the extinct population of a remote period of our globe,—those enormous pachydermata, the mastodons and mammoths, that lie buried in the alluvial and superficial strata. We now arrive at the consideration of phenomena not less interesting—the occurrence of the skeletons of immense numbers of carnivorous animals in fissures and caverns. In the former discourse I alluded to the cavities which abound in certain rocks of limestone, and described the process by which their roofs, floors, and walls were coated with sparry incrustations, and ornamented with stalactites and stalagmites. Some of these caverns appear to have been occasioned by the destruction of the softer portions of the rock by subterranean streams; others are so extensive, and present such decided marks of angular fracture, as to leave no doubt that they have been produced by the shocks of earthquakes. The occasional occurrence of the bones of animals in such cavities might reasonably be expected. Those that admitted of easy access from without, might be frequented by those animals whose habits lead them to retire into dark and secret recesses; while other species, as kids, deer, &c. might fall

into open fissures, and their bones thus become
enveloped and preserved in calcareous incrustations.   But the immense quantities of only one
or two species of carnivora that are found in some
caverns, show that these have been for a long period
the dens of extinct species of bears, wolves, tigers,
hyenas, and other carnivorous tribes.

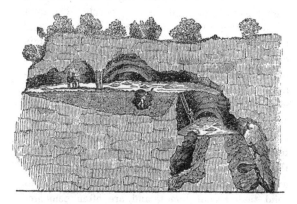

TAB. 17.—THE CAVES OF GAYLENREUTH.

29. CAVES OF GAYLENREUTH.—For many centuries, certain caverns in Germany have been celebrated for their osseous treasures, particularly those
in Franconia : the most remarkable is that of
Gaylenreuth, which lies to the north-west of the
village, on the left bank of the river Wiesent.   The
entrance, which is about seven feet high, is in the

face of a perpendicular rock, and leads to a series
of chambers from fifteen to twenty feet high, and
several hundred feet in extent, terminated by a
deep chasm, which, however, has not escaped the
ravages of visitors. This cavern is perfectly dark,
and the icicles, or pillars of stalactite, reflected by
the torches which it is necessary to use, present a
highly picturesque and striking effect. The floor
is literally paved with bones and fossil teeth; and
the pillars and corbels of stalactite also contain
osseous remains. Loose animal earth abounding in
bones, forms in some parts a layer ten feet in thickness. A highly graphic description of this cave
was published by M. Esper, more than sixty years
ago; at that period, some of the innermost recesses
of the cave contained waggon loads of bones and
teeth; some imbedded in the rock, and others in
the loose earth. The bones in general are scattered
and broken, not rolled; they are lighter and less
solid than recent bones, and are often cemented
with stalactite. Through the kindness of Lord
Cole, and Sir Philip Egerton, I am able to illustrate
these remarks by a very extensive suite of osseous
remains, exhumed from the deepest recess in the
cavern; and which were collected a short time
since by these distinguished geologists. But the
most interesting specimen in my possession is this
remarkably perfect skull of a Bear, which belonged
to my friend the late Mr. Parkinson, the author of
that delightful work, The Organic Remains of a

Former World. A comparison of this relic with the skull of the Polar bear, shows that it must have belonged to a species of that animal.* Cuvier, who enjoyed the opportunity of examining a very large collection of bones from Gaylenreuth, was enabled to determine that at least three-fourths of the osseous contents of the caverns belonged to some species of bear; and the remaining portion to hyenas, tigers, wolves, foxes, gluttons, weasels, and other small carnivora. By the bones which were referrible to the bear, he established three extinct species; the largest of these has a more prominent forehead than any living species, and is called the *Ursus spelæus*, or bear of the caverns, and it is to this species the skull I have just exhibited belongs; the other has a flatter forehead, and has been named *Ursus arctoidæus*. The hyena was allied to the spotted hyena of the Capes, but differed in the form of its teeth and head. Bones of the elephant and rhinoceros are also said to have been discovered; together with those of existing animals, and fragments of sepulchral urns of high antiquity.

30. FORSTER'S HÖHLE.—Another cavern in this part of Germany is mentioned by Dr. Buckland, as one of the most remarkable for the beauty of its

* Their Royal Highnesses the Princes George of Cumberland and Cambridge, when inspecting my collection a few years since, at Lewes, pointed out this skull to me as resembling some fossil remains that had been exhumed from a fissure in limestone, in the kingdom of Hanover.

incrustations. Its height varies from ten to thirty feet, and its greatest width is about ten yards; in the side vaults, or recesses, which descend, at an angle of about forty-five degrees, into the main chamber, the stalagmite has formed the appearance of cascades of pure alabaster, the waves of which seem to be rushing out at the bottom, to pour themselves into the stagnant lake of the same substance which covers the floor. The rocky roof has been corroded into deep cavities, which are separated by partitions of every conceivable form and tenuity, giving it the appearance of the richly fretted Gothic roof of a chapel, with pendent corbels. Beautiful stalactites depending from these projections, reach almost to the floor, and contribute by their delicacy and transparency to give additional richness to the scene.

It is certainly, as M. Cuvier remarks, a most extraordinary fact, that caves, spread over an extent of two hundred leagues, should have the same osseous contents. The relative proportions of the different species are computed to be as follow :— three-fourths of bears—two-thirds of the remainder of hyenas—and a small number of the tiger or lion, and of the wolf or dog; rolled pebbles of a greyish blue marble are the only extraneous materials found with the bones. Let me here call your attention to the singular association of species which some of these caves present; their recent types being widely separated. Thus in one cavern, animals allied to the spotted hyena of the Cape of Good Hope, are collocated with the remains of others related to the glutton, which inhabits Lapland; and in another, bones of the rhinoceros are associated with those of the rein-deer. Numerous caves containing similar remains are scattered over the continents of Europe and America; and even in Australia, bones are found in caverns.

31. BONE CAVERNS IN ENGLAND—KIRKDALE CAVE.—Several caverns containing bones of bears, and other carnivora, in every respect analogous to those of Germany which we have just described, have been discovered and explored. Dr. Buckland, in his valuable work, the *Reliquiæ Diluvianæ*, has noticed several of the most interesting assemblages of this kind. The cave of Kirkdale, now so well known in consequence of the highly interesting disquisition on its contents by my distinguished friend, is one of the most celebrated; and it is highly gratifying to me, that I can illustrate these remarks by an extensive series of bones, &c. from that cavern. In the summer of 1821, a cave was discovered near Kirkdale, about twenty-five miles NN.E. of York, in a bank about sixty feet above the level of a small valley, and near a public road. Some workmen who were quarrying stone, cut across the narrow mouth of a chasm, which had been choked up with rubbish, and overgrown with grass and bushes; and which from this cause, as well as from its inaccessible situation, had hitherto

escaped observation, the entrance being so small
that it was only possible for a person to enter in a
bent position. The passage is exceedingly irregular
in its dimensions, varying from two to seven feet in
breadth, and from two to fourteen feet in height ;
its greatest length is 245 feet. It divides into
several smaller passages, which have not yet been
explored, as they are nearly closed by stalactital
concretions ; these cavities occur where the roof is
intersected by fissures, which are continued for a
few feet, but are gradually lost in the superincumbent limestone, and are thickly lined with stalactites.
The true floor of the cave is only seen near the
entrance ; for in the interior the whole has been
covered with a bed of hardened mud or clay, about
a foot in average thickness ; the surface of which
was perfectly smooth and level when the cave was
first opened, except where stalagmites had formed
upon it by infiltration from the roof. Where stalactitic matter incrusted the sides, it also extended
over the bottom like a thin coat of ice ; and therefore must have been formed since the mud was
introduced. This mud or clay was filled with fragments of bones belonging to a great variety of
animals ; and some of the bones exhibited marks of
having been gnawed. From many corroborative
circumstances these appearances are, with much
probability, supposed to have been occasioned by
hyenas. The bones thus preyed upon belong to
the tiger, bear, wolf, fox, weasel, elephant, rhino-

ceros, hippopotamus, horse, ox, and deer.   Bones
of a species of hare or rabbit, water-rat, and mouse,
with fragments of the skeletons of ravens, pigeons,
larks, and ducks, were also discovered with these
remains.

TAB. 18.—JAW OF A HYENA, FROM KIRKDALE CAVE.*

From these facts Dr. Buckland infers that the
cave at Kirkdale was inhabited by hyenas for a
considerable period, and that many of the remains
found there were of species which had been carried
in, and devoured by those animals ; and that in
some instances the hyenas preyed upon each other.
Portions of elephants' bones seem to show that occasionally the large Mammalia also served as food ;
but it seems probable that many of the smaller
animals may have been drifted in by currents, or
have fallen into the chasm, through fissures now
closed up by stalactite.

KENT'S CAVE, near Torquay, which is nearly

* This figure is from Dr. Buckland's representation of a
portion of the lower jaw of a hyena, from Kirkdale Cave.

600 feet in length, has yielded immense quantities
of bones of carnivora; and in the Isle of Portland,
at Plymouth, and in the Mendip Hills, similar
accumulations have been found. In the south-east
of England but one instance is known; a fissure in
the sand-rock at Boughton Quarries, near Maidstone, contained the jaw and bones of a hyena,
which are now in Dr. Buckland's Museum, at
Oxford. This fact is highly interesting to us, for
it proves the existence of the same condition of
animated nature in this part of our island, as in the
districts previously mentioned; and I cannot doubt
that sooner or later bone caverns will be found in
the south-east of England. Very recently a cavern
has been discovered near Plymouth, in which hyenas'
bones were in abundance associated with those of
the elephant, rhinoceros, horse, &c.

From what has been stated, we learn that our
wastes and forests were once inhabited by carnivorous animals of extinct species and genera, which
are now confined to southern climates; — that
these lived and died for successive generations, and
were the prey or the devourers of each other;
— that the hyenas, according to their peculiar
habits, dragged into their dens the creatures which
they killed or found dead, and devoured them at
their leisure; — that subsequently the races were
annihilated, and have been succeeded by animals
altogether of a different character.

32. DISEASED BONES OF CARNIVORA FOUND IN

CAVES.—Among the bones found in the caves of
Germany are many in a condition which must have
resulted from accident or disease. In some there has
been a formation of new bony matter to repair fractures; in others there is anchylosis, or adhesion of
the joints from inflammation: while in some the
effects of caries, or decay of the bones, the result of
tedious and painful diseases, are apparent. Others
have a light and spongy character, and are very
fragile, which must have arisen from a want of
energy in the nutritive system, in consequence of
a scrofulous affection.*

33. HUMAN BONES, AND WORKS OF ART, IN
CAVERNS.—Bones of man, and fragments of ancient
pottery, have been found in caves, both in France
and Germany; a circumstance perfectly natural,
since we are well aware that mankind, in a rude
state, have been in the habit of living in caves, and
traces of their having inhabited recesses, which
had previously been the retreat of wild animals,
were therefore to be expected. But as bones of extinct animals occurred with them, it was rashly assumed that they were coeval with each other; more
accurate observations have, however, shown that the
human remains were introduced at a subsequent
period. We have historical proof that the early inhabitants of Europe often resided, or sought shelter

* Professor Walther, on the Antiquities of Diseases of Bones;
see Professor Jameson's Cuvier's Theory of the Earth. Edin.
1827.

in natural caves. Thus Florus relates, that Cæsar ordered the inhabitants of Aquitania to be enclosed in the caverns to which they had retired. Many tribes of the Celtic race occupied these subterranean retreats, not only as a refuge in time of war, but also for shelter from cold ; as magazines for their corn, and for the produce of the chase ; and as places of concealment for the animals which they had domesticated. The bones of such of these people as perished, or were buried in these caverns, would become blended with the mud, gravel, and debris of the animals already entombed ; and a stalagmitical paste would in some places be formed by the infiltration of water, as at Bize, and would thus cement the whole into solid aggregates. We should therefore expect to find masses of stone, containing bones of the bear, and other extinct species, with human bones, fragments of pottery, and terrestrial shells, and bones of animals of modern times. Such are the contents of numerous caves, and the above explanation points out the mode in which these accumulations have taken place.*

34. Osseous Breccias, or Bone Conglomerates.—The phenomena we have next to examine are even more extraordinary than those which have already been described ; for the osseous remains which now claim our attention are not imbedded in gravel or clay, or collected together in caves, but

* Memoir by M. Desnoyer.

are found in fissures of limestone, which extend over
an area of many hundred leagues, and occur in detached rocks and islands, very remote from each
other. The limestone presents but little variety, and
the substance in which the bones are enveloped
is every where the same; the fossil remains belonging, with but few local exceptions, to similar
species of animals. The rocks are split in every
direction, and the fissures filled with what geologists term an osseous or bone breccia; that is,
bones, and fragments of bones, held together by a
calcareous cement or paste; in the same manner as
the conglomerated shingle of Kemp Town; or, to
exemplify its nature by a still more familiar illustration, the mixture of mortar, pebbles, &c. employed in masonry, and called concrete. This
cement is of a reddish-brown, very much resembling
common brick; and the bones are beautifully white,
having in many instances their cavities lined with
spar, as in these specimens from Gibraltar. In
some examples the bones have undergone but little
change; in others, the cells in the cancellated structure, are filled with calcareous matter; as you
may observe in this specimen from Cerigo, (presented to me by Lady Mantell,) which is cut and
polished, to show the internal structure of the
bones. This tooth of a species of ruminant (Tab.
11, p. 119,) from Gibraltar, resembles in its general
appearance the teeth found in the Coombe-rock of
the Brighton cliffs. But the stone to which it is

M

attached is more compact, and partakes of the character of marble; it is of a dull red colour, mottled with white, and is susceptible of a good polish. This osseous breccia occurs on the northern shores of the Mediterranean; in the rock of Gibraltar; at Cette, Nice, and Antibes; in Dalmatia, and in the isles of Cerigo, Corsica, &c.; and in Sicily, and Sardinia, and many parts of Germany. Each of these localities present highly interesting examples of the immediate subject of our inquiry.

35. THE ROCK OF GIBRALTAR. — The rock of Gibraltar, so well known from its historical and political importance, affords an admirable illustration of the phenomena under review; and, for the sake of brevity, I shall chiefly confine my observations to that celebrated spot. Gibraltar is situated, as you well know, on the Spanish side of the Mediterranean, and is united to the main land by a narrow isthmus, which is about three-fourths of a mile broad, eight or ten feet above the level of the sea, and formed of consolidated sand. The rock stands on the western extremity of the area in which the osseous breccias occur, and its greatest altitude is about 1,350 feet. It consists principally of a compact, bluish-grey marble, which, like most limestone masses, is cavernous. The principal cavern is called St. Michael's, and contains stalagmites and stalactites, some of which when polished are of great beauty. In the fissures with which the rock

is intersected, as well as in some of the caves, a calcareous concretion, of a reddish-brown colour, occurs, which in some parts is a mere earthy mass, but in others is highly indurated. The bones are commonly in a broken state, and but seldom water-worn; and the fragments of limestone, with which the fissures abound, are also angular, and have evidently, like the bones, fallen into the crevices at different periods, and been gradually incrusted and conglomerated by calcareous infiltrations. Snails and other land shells are often found impacted in the solid breccia; they belong to the existing species of the country. As the concretion is still in progress of formation, masses may be found in which terrestrial shells occur, unmixed with bones. The cementing material is very similar in the different localities where the breccia has been observed; namely, at Cette, Nice, Antibes; in Dalmatia and Sardinia. The animal remains of the breccia are referrible to several species, some recent and others extinct, of deer, antelope, rabbit, rat, mouse, &c. Bones of birds and of lizards have been found at Cette; and of lemmings,* and of the *Lagomys*,† which now only exists in Siberia: it is but rarely that traces of carnivora are observed. No one can fail to be struck with surprise at the occurrence of these isolated, yet analogous phenomena, which

* Lemming, or Lapland marmot.

† Signifying rat-hare. A genus of animals which forms a link between the hare and the rat.

surround the great basin of the Mediterranean—
rocks of a uniform character, fissured and broken,
their rents filled up with similar materials, and with
the remains of the same species of animals. The
occurrence of species, either extinct, or no longer
inhabiting the same latitudes (as the *Lagomys*),
refers the period of the existence of these animals to
the epoch of the mammoths and mastodons; and the
absence of marine remains, and of the usual abrad-
ing effects of water, show that the breccia was
formed on dry land, and not beneath the waters of
the ocean.

The rational explanation of these phenomena
appears to be that which assumes the original union
of these distant rocks and islands into a continent,
or large island, which, like Calabria, was subject to
repeated visitations of earthquakes; and that the
animals which inhabited the country fell into the
fissures thus produced, and were preserved by the
calcareous infiltrations that were constantly in progress. Subsequent convulsions severed the country
into rocks and insular masses, of which catastrophe
the osseous conglomerates are the physical and only
record.

36. OSSEOUS BRECCIA IN AUSTRALIA.—Caves
and fissures, filled with osseous breccia, in the same
manner with those I have described, have been
discovered also in New Holland, to the westward of
Sydney, near the banks of the Macquarrie river;
and it is not a little remarkable, that even the red

ochreous colour of the European conglomerate prevails; the bones, however, belong to animals wholly distinct from any hitherto noticed in the preceding examples. Some of them are of living, others of extinct species, but all of them are referrible to marsupial animals, as the *Kangaroo, Wombat, Dasyurus,* &c. A portion of a large bone, found in a cave, is said to resemble the leg-bone of the hippopotamus, but this requires confirmation; it is, however, a subject worthy of attention, since the kangaroo is the largest animal now known in those regions. The fact that all the fossil animals of Australia are marsupial, that is, belong to the creatures which carry their young in a pouch—a type of organization which is the peculiar feature of the existing races of the country—is also of great interest; for it proves the marsupial character of the zoology of those regions from a very remote period.

37. RETROSPECT.—I must now bring to a close this examination of the ancient superficial deposites —those accumulations of alluvial matter, which, taken as a whole, are referrible to a more early period than those which strictly belong to the modern or human epoch. And as in the former discourse I found it necessary to dwell on the discoveries of astronomy, to elucidate some of the physical changes of our planet; in the present I have summoned comparative anatomy to our aid, and have endeavoured to point out the mode of

induction pursued by the palæontologist, in his
inquiries into the fossil remains of animal organization, by which he is enabled to call forth from
their rocky sepulchre the beings of past ages, and,
like the fabled sorcerer, give life and animation to
the inhabitants of the tomb. From the facts that
have been presented to us in the course of this lecture, we arrive at the following important inferences :—

First, That the extinction of certain forms of
animal existence is a law, which is not only in
operation at the present moment, but has extended throughout the period comprehended in
our present researches; and we have traced its
influence from the partial extirpation of certain
existing species, to the entire annihilation of species
and genera that once were cotemporary with man ;
as well as to those which are known to have lived,
and become extinct, long prior to the creation of
our race.

Secondly, That while in the modern marine and
fluviatile accumulations, the remains of existing
species of animals, and of man and his works, are
entombed, in the ancient deposites of water-worn
materials, those of large mammalia alone are imbedded.

Thirdly, That the animal remains principally
belong to extinct Pachydermata, related to the
elephant, hippopotamus, and sloth; with horses,
deer, and other ruminants ; and that these had for

their contemporaries bears, hyenas, tigers, and other carnivora belonging to extinct species.

Fourthly, That there was therefore a period immediately preceding the existence of man, when the earth teemed with large herbivorous animals, which roamed through the primeval forests unmolested, save by beasts of prey. Numerous species and entire genera have been swept away from the face of the earth, — some by sudden revolutions, others by a gradual extinction, — while many have been exterminated by man.

Lastly, That these various deposites, whether formed in the beds of lakes or rivers, or in the estuaries and basins of the ocean, have been elevated above the levels of the waters, and now constitute fertile countries, supporting the busy population of the human race.

I have thus endeavoured to interpret one page of the ancient physical history of our planet, and to explain the records of one epoch in geological chronology. We have entered upon the confines of the past, and already we find ourselves surrounded by an innumerable population of unknown types of being — not as dim and shadowy phantoms of the imagination, — but in all the reality of form and structure, and bearing the impress of the mighty changes of which they constitute the imperishable memorials. We have again witnessed the effects of the continual mutations of the land and water,—

have seen that our present plains and valleys were submerged beneath the ocean, at a period when large mammalia, apparently unrestricted by existing limits of climate, were inhabitants of regions which are now no more—and we have obtained additional proof that—

> New worlds are still emerging from the deep,
> The old descending in their turn to rise!

Even in this early stage of our progress, we have conclusive evidence of the extinction of whole tribes of animals, equally admirable in their adaptation to the condition in which they were placed, as the races which still survive. And delightful it is to the geologist, to find that this fact, which but a few years since was received with hesitation by most, and condemned and rejected by many, is now adduced by the moralist and the divine, as affording fresh proofs of the wisdom and overruling providence of the Eternal. Reflecting on these phenomena, the mind recalls the impressive exclamation of the poet—

> My heart is awed within me, when I think
> Of the great miracle which still goes on
> In silence round me—the perpetual work
> Of THY creation, finished, yet renewed
> For ever!

LECTURE III.

1. INTRODUCTORY REMARKS. — It is my object in these Lectures to present a general view of the philosophy of Geology, rather than enter at length on the nature and distribution of the materials of which the crust of our globe is composed, —and to render the details of geological phenomena subservient to an explanation of the laws which the Divine Author of all things has established for the production, maintenance, and government of

the organic and inorganic kingdoms of Nature. Based as Geology is upon observations of the various physical changes which are now taking place, and on inquiries into the natural records of those changes, in periods antecedent to all human history and tradition, the rocks and mountains are the alphabet—the book of Nature the volume—by which the student of this interesting department of science can best learn its important lessons. But to those who cannot examine Nature in her secret recesses, or accompany an experienced teacher to the valleys, or to the mountain-tops, lectures, illustrated by specimens and drawings, afford, perhaps, the best substitute for the more efficient and delightful mode of instruction.

That we may obtain a clear and comprehensive view of the vast field of inquiry that lies open before us, artificial classifications are necessary, in this as in other departments of science; and without assuming that the arrangement, in which the various deposites are grouped by geologists, will not, in the progress of discovery, require considerable modification, I purpose, as an introduction to the subjects hereafter to be discussed, to place before you a tabular view of the rocks and strata in their presumed chronological order. At the same time it is necessary to bear in mind, that all classifications of this kind must necessarily involve arbitrary distinctions, and that very possibly it will hereafter be found that we may in some instances have classed as general, what may prove to be merely local phenomena; and grouped together rocks and deposites, which farther investigations may show to be distinct, and separated from each other by vast periods of time. But this consideration will not effect those leading principles of modern Geology, which it is my present endeavour to render familiar to the intelligent but unscientific inquirer.

We will now take a general view of the nature of the mineral substances which enter into the composition of the crust of our globe ; and briefly notice the laws which regulate the deposition of detritus in the beds of the lakes and rivers, and in the depths of the ocean. But as I have already remarked, it is not my intention to enter on these departments of Geology in detail; the works of Mr. Bakewell,* Lyell,† Phillips,‡ Delabeche, § and

* Introduction to Geology, by Robert Bakewell, Esq. 1 vol. 8vo. This excellent volume should be the first book in the library of the young geologist.

† The Principles of Geology, by Charles Lyell, Esq. F.R.S. 4 vols. 12mo. One of the most interesting works in the English language. I have reason to believe that an elementary work will shortly be published by my distinguished friend, that will supply what has long been a desideratum—a popular view of the " Elements of Geology."

‡ A Guide to Geology, by John Phillips, Esq. F.R.S. 1 vol. 12mo. The article on Geology, in the Encyclopedia Metropolitana, by the same excellent writer, is in my opinion the best scientific epitome of modern Geology that has yet appeared.

§ A Geological Manual, by H. T. De la Beche, Esq. F.R.S. An admirable work of reference.

others, afford every information on these subjects
which the student can require.

2. SUBSTANCES COMPOSING THE CRUST OF THE
GLOBE.—Every substance is composed of atoms of
inconceivable minuteness, held together by a principle which is termed attraction or cohesion, and is
probably a modification of that influence, which,
as it exists under other conditions, in inorganic
substances, is called electricity, galvanism, or mag-
netism; and in organized beings, nervous influence.
As the different states of solidity, fluidity, or vapour
in which every material body may exist, have been
exemplified in the former lectures, we now need only
remark, that there are about sixteen substances,
which in the present state of chemical knowledge
are considered simple in themselves, and that these
constitute by their various combinations, if not entirely, by far the largest amount of the matter,
either gaseous, liquid, solid, organic, and inorganic,
of the earth.   Of these, eight are non-metallic
substances ; viz. *oxygen, hydrogen, nitrogen, carbon, sulphur, chlorine, fluorine,* and *phosphorus.*
There are also six metallic bases of alkalies and
earths, namely, *silicium, alumine, potassium, sodium,
magnesium,* and *calcium ;* and two, the oxides of
which are neither earths nor alkalies, namely, *iron*
and *manganese.*   The remaining metallic substances,
copper, lead, zinc, arsenic, silver, gold, &c. are comparatively unimportant in a geological point of view.*

* De la Beche.

The common sedimentary rocks are in a great
degree composed of lime, silex, or argillaceous
earth. These substances have, what in mineralogical language is called, a *cleavage*, or peculiar
fracture, which is distinct in each. Thus, if I take
a flint and break it at random, you perceive that it
still preserves a glassy or *conchoidal* fracture, a sharp
cutting edge; and subdivide it as I may, it still
retains the same character: but if I break a piece
of chalk, the edge is not sharp or cutting, but blunt
and dull, exhibiting what is called in mineralogy an
earthy fracture. Again, if I shiver to pieces with my
hammer this calcareous spar, every fragment presents, more or less distinctly, a rhomboïdal form;
so true is the remark, that we cannot break a stone
but in one of nature's joinings.

3. CRYSTALLIZATION.—Crystallization may be
defined a methodical arrangement of the particles
of matter according to fixed laws. For instance,
there are nearly 500 varieties of crystallized carbonate of lime, each crystal being made up of
millions of atoms of the same compound substances,
and having one invariable primary form—that of
a rhomboid. Mechanical division is incapable of
altering this arrangement; break them as we may,
we can only separate them into a rhomboidal figure;
nor can this condition be altered except by chemical
decomposition. If we pursue our investigations
yet farther, analysis shows that every atom of these
crystals is composed of quicklime and carbonic

acid, which are each made up of countless molecules. "Lime and carbonic acid are also themselves
compounds, lime being composed of a metal called
calcium and oxygen ; and carbonic acid, of carbon
and oxygen. Thus these ultimate particles of calcium, carbon, and oxygen, form the indivisible atoms
into which all the secondary crystals of lime may be
reduced."*

4. STRATIFICATION.—As our previous investigations have shown that the disintegration and
solution of the most refractory, and apparently in-
destructible substances, by the conjoined effects of
mechanical and chemical agency, are constantly in
progress, we can at once proceed to the consideration of the manner in which the spoils of the
ancient land have been accumulated, and converted
into the rocks and strata of existing islands and
continents. We have already adverted to the formation of beach and sand, and the deposition of
mud and clay in layers or strata, and their subsequent
consolidation into rocks. And here let me remind
you, that *strata*, are the successive layers or accumulations of detritus, spread over each other, in such
manner as to allow of the partial consolidation of
one bed, before it is covered by a deposition of the
materials of another; and a rock is said to be *stratified*, when it presents the appearance of such divisions. The chalk cliffs, and the sandstone quarries

* See Dr. Buckland's Bridgwater Essay.

in the South-east of England, afford excellent illustrations of this structure. The original direction of these layers must have been more or less horizontal, for this obvious reason, that from their fluid, or semi-fluid state, they would find their own level, and spread over the surface of the basin into which they flowed; and although they might partake of the inequalities of the depression in which they were deposited, yet this cause would not affect their general distribution. The strata when accumulated in very thin layers, resembling the seams formed by the leaves of a closed book, are termed *laminæ ;* and this character very commonly prevails in fluviatile or river deposites: thus the shells, aud clays, and sandstones, in Tilgate Forest are laminated, and often bear the impress of the waters which have meandered over them. (See page 36.) The contemporaneous deposites formed in the same oceanic basin, however they may maintain a general character over very extensive areas, must nevertheless vary considerably. At the present moment, the rivers flowing from different latitudes into our existing seas, must necessarily be producing in the same marine basin accumulations of a very dissimilar character; and the geographical distribution of the detritus, must be still more affected by the agency of those mighty currents, to which allusion has already been made (page 47). Bearing in mind these elements of variation in the depositions that may contemporaneously take place within

the same hydrographical basin, we shall be prepared
to find similar discrepancies in the contents of the
beds of the ancient oceans.

5. INCLINED AND VERTICAL STRATA.—But
although the strata, whether accumulated in banks
or ridges, or deposited in basins or depressions,
have originally been consolidated in horizontal
layers, yet this arrangement has frequently been
disturbed by expansive forces from below, and the
strata have been broken up, and thrown into every
direction, from a slight degree of inclination, to a
vertical position. The sections before us, (Plates
5, 6,) to which I shall hereafter have occasion
to refer, exhibit strata in various states of displacement.

Although it is my wish to abstain as much as possible from technical language, yet in some instances
its use cannot be avoided without much circum-
locution ; I will therefore explain a few convenient
terms :—Thus, when a series of strata present
the same direction, like books piled horizontally
upon each other, (Plate 5, No. II. *Section near
Devizes,*) they are said to be *conformable ;* but when
beds are superimposed on others that lie in a different direction, (Plate 5, No. III. *The fresh-water
limestone, &c. on the Lias,*) as if a set of horizontal
volumes were placed flat on the inclined edges of
another series of books, they are, in geological language, in an *unconformable* position.

6. VEINS AND FAULTS.—But not only have the

strata suffered change of position from the disturbing causes which we have seen are in continual
action; they have also been rent and broken up,
and exhibit cracks or fissures, which in rocks near
the surface are sometimes filled, as we have already
noticed, with bones, pebbles, and stalactitical concretions (page 160); and in those of ancient epochs,
with eruptions of melted matter, and metalliferous ores. The term *fault* is applied to those
fractures and displacements of the strata which are
accompanied with the subsidence of one part of a
mass and the elevation of another. This is exemplified in this section of the carboniferous strata,
(plate 3, fig. 7,) where three curved layers, or seams
of coal, have shifted to a lower level, although both
sides of the rock remain in apposition. In short,
stratification may be compared with the operations
of man in erecting various buildings; strata of clay
being comparable to beds of mortar, those of harder
rocks to layers of brick; while the fissures, veins,
and faults are analogous to the cracks, sinkings,
and displacements produced by the settling of various portions of the whole erection.

7. CHRONOLOGICAL ARRANGEMENT OF STRATA.
—In the ancient alluvial gravel, sand, and marl, containing the remains of gigantic mammalia, which
formed the principal subject of the last lecture, but
few indications of stratification occur; these deposites, for the most part, bearing the character of
materials transported by river currents, or accumu-

N

lated in estuaries, and thrown up in bays and creeks by the waves, rather than of tranquil depositions. In the formations which succeed, we shall find whole countries composed of regularly stratified rocks, interspersed here and there with diluvial debris. The plan of the strata before you (plate 3) is intended to present a general view of the various systems of rocks, from the modern to the most ancient, as classified by modern geologists. For more detailed explanations and sections, I refer you to the works already noticed, and particularly to the admirable systematic diagram of that veteran geologist, Mr. Webster, which forms the frontispiece of Dr. Buckland's Bridgwater Essay.

CHRONOLOGICAL ARRANGEMENT OF THE STRATA.

Commencing with the uppermost or newest Deposites.

FOSSILIFEROUS STRATA.

1. MODERN AND ANCIENT ALLUVIUM.—Comprising the modern and ancient superficial deposites, described in the previous lectures. The *modern* are characterized by the remains of man and contemporaneous animals and plants ; the *ancient*, by an immense proportion of large mammalia and carnivora, of species and genera, both recent and extinct.

II. THE TERTIARY STRATA.—An extensive series, comprising groups of marine and lacustrine deposites, characterized by the remains of animals and vegetables, the greater portion of which are extinct. Volcanoes of great extent were in activity during this epoch.

III. THE CHALK, or CRETACEOUS SYSTEM.—A marine formation, comprising beds of limestone, sandstone, marl, and
clays, abounding in remains of zoophytes, mollusca, cephalopoda, echinodermata, fishes, &c. ; drifted wood, and marine
plants, with crocodiles, turtles, and extinct reptiles.

IV. THE WEALDEN.—Comprising the Weald clay, the strata
of Tilgate Forest, and the limestones and clays of Purbeck.
A freshwater formation, evidently the delta of some ancient
river; characterized by an abundance of the remains of
enormous and peculiar reptiles, the Iguanodon, Hylæosaurus,
Megalosaurus, Plesiosaurus, Crocodile, &c., of terrestrial
plants, freshwater mollusca, and *Birds*.

V. THE OOLITE.—A marine formation of vast extent, consisting of limestones and clays, abounding in marine shells,
corals, fishes, and reptiles, both terrestrial and marine.
Land plants of peculiar species, *and the remains of one genus
of* MAMMALIA.

VI. THE LIAS. — A series of clays, shales, and limestones,
with marine shells, cephalopoda, crinoidea, and fishes.
Reptiles, particularly of two extinct genera, the Plesiosaurus and Ichthyosaurus, in immense quantities. Drifted
wood, and plants.

VII. THE SALIFEROUS STRATA.—Comprising marls, sandstones,
and conglomerates, frequently of a red colour, with shells,
corals, and plants ; fishes and reptiles. This series forms
the grand depository of rock-salt.

VIII. THE CARBONIFEROUS SYSTEM, or COAL.—Shales, iron-
stones, millstone grit, freshwater limestone, and immense
beds of coal. The lower part of the series consists of red
sandstone and conglomerates. This system is characterized
by innumerable remains of land and aquatic plants of lost
species and genera, and of a tropical character; fishes and
reptiles ; insects.

IX. THE SILURIAN SYSTEM.—Composed of marine limestone,

shales, sandstones, and calcareous flags; abounding in shells, many of new forms; and swarming with corals, crinoidea and trilobites.

X. THE CUMBRIAN, or GRAUWACKE SYSTEM.—Consists principally of a largely developed series of slate rocks, and conglomerates, with shells, and corals.

METAMORPHIC ROCKS.

Destitute of Organic Remains.

Stratified.

XI. THE MICA SCHIST.—Sedimentary rocks, altered by high temperature; mica slate, quartz rock, crystalline limestone, gneiss, and hornblende schist, &c., exhibiting no traces of organic remains.

XII. THE GNEISS SYSTEM.—Formed of gneiss, sienite, and quartz rock, alternating with clay, slate, mica, schist, &c.

Unstratified.

XIII. GRANITE. In amorphous masses and veins; porphyry, serpentine, trap, &c.

I have represented in this diagram (plate 3) intrusions of the ancient melted rocks, as serpentine, porphyry, trap, and granite, into the sedimentary strata; and metalliferous and granitic veins in the granite, to which I shall hereafter have occasion to refer. At present it will only be necessary to mention, that the leading features of this arrangement may be recognised in every considerable extent of country throughout the world; but the lesser divisions are more local, and cannot be generally maintained, for reasons which must be sufficiently obvious, after what has already been advanced.

From this general view of the physical records of the mutations which the surface of our globe has undergone, we learn how numerous and important are the phenomena comprised within the sphere of geological inquiry, and how vain is the attempt to offer more than an epitome of its wonders in the brief space allotted to a popular course of lectures. We will now enter upon the consideration of the TERTIARY FORMATIONS, those deposites of the seas, and rivers, and lakes, which are referrible to the period immediately antecedent to the existence of the mammoth and mastodon, and subsequent to the deposition and consolidation of the chalk.

8. TERTIARY FORMATIONS. — The important discoveries of MM. Cuvier and Brongniart, about twenty years since, in the immediate vicinity of Paris, first directed the attention of geologists to the important series of deposites which are now distinguished by the name of tertiary (see p. 14). The fossil bones which abound in the gypsum quarries of Montmartre, and belong to extinct genera of mammalia, were by the genius of Cuvier, again called into existence, and the philosophers of Europe saw with astonishment, whole genera of unknown and extraordinary types of being, disinterred from rocks and mountains, which had hitherto been considered as possessing no scientific interest. The existence of analogous strata, some marine, others of a lacustrine and fluviatile character, has since been discovered throughout the continents of Europe

and America, forming series so vast and extensive, and requiring such a lapse of time for their
production, that the chalk, hitherto considered as
comparatively modern, is carried back to a period
of immense geological antiquity. The tertiary
system may be said to constitute a series of formations which link together the present and the
past; for while the most ancient contain organic
remains related to the secondary formations, the
most recent insensibly glide into the modern deposites, and contain many existing species of animals
and plants, associated with forms that are now
blotted out for ever. Mr. Lyell has formed a classification of the tertiary strata, founded on the proportion of recent species of animals which they contain;
and as shells occur in many of the strata in great
abundance, and in an excellent state of preservation,
he has selected those types of animal organization
for the distinctive characters of the subdivisions
into which, for the convenience of study, he pro-
poses separating these deposites. In the present
state of our knowledge, this arrangement is of great
utility, but it appears probable that it may require
considerable modification, or, perhaps, hereafter be
altogether abandoned with the progress of geological researches; for it cannot be concealed, that
strata in which no recent species have yet been
found, may yield them to more accurate and extended observations.

9. CLASSIFICATION OF THE TERTIARY STRATA.

—Mr. Lyell has arranged the tertiary system into four principal groups, each characterized by the relative proportion of recent and extinct species of shells which they contain, and he has adopted a nomenclature denoting the characters upon which this arrangement is founded. They are as follow :—

THE PLIOCENE (signifying more new or recent). —Tertiary strata, in which the shells are for the most part recent, with about ten per cent of extinct species ; these beds are again subdivided into the newer and older pliocene.

THE MIOCENE (denoting less recent).—Containing a small proportion, about twenty per cent., of recent species.

EOCENE (signifying the dawn of recent, in allusion to the first appearance of recent species).— Containing very few recent species, perhaps not more than three or four per cent.

The marine formations are associated with a like number of freshwater deposites, and the general characters of the tertiary system are alternations of marine and lacustrine strata. The districts occupied by these deposites in Europe, are exceedingly variable in extent, as Mr. Lyell has shown in a very ingenious map of the tertiary seas ;* and it appears certain, that during the epoch of their formation, there were areas which were alternately the sites of freshwater lakes and inland seas, and that these

* Mr. Lyell's Principles of Geology, vol. i. p. 214.

changes were dependent on oscillations of the relative level of the land and water.

10. Fossil Shells.—The geological evidence
afforded by the remains of animals and plants has
already been fully exemplified; but our remarks
have hitherto in a great measure been confined to
the fossilized skeletons of terrestrial quadrupeds;
shells, however, from their durability, often escape
obliteration under circumstances in which all traces
of the higher orders of animals are lost. In loose
sandy strata, they occur in a high degree of perfection; in mud and clay, in a fragile state; in some
instances they are silicified; and many limestones
are wholly composed of these remains, cemented
together by calcareous matter. Molluscous animals* are divided into *Mollusca*, properly so called,
which are covered with a shell, as snails, periwinkles, &c.; and *Conchifera*,† having a shell with
two valves, as the oyster, scallop, &c. The former
are of a higher organization than the latter, having
eyes, and consequently a distinct nervous system; the
latter have neither eyes nor head, and are therefore
called *Acephala*.‡ Like terrestrial animals, some
genera of mollusca are herbivorous, living exclusively on vegetables; and others are carnivorous,
having commonly a retractile proboscis, by which
they can perforate wood, shells, and other substances.
This instrument is protected by a canal, and the

* Soft bodied animals.   † Shell-bearing animals.
‡ Having no head.

shells of the carnivorous testacea are, therefore, generally provided with a channel or groove for the passage of that organ (as in Tab. 22, Fig. 3, 4, 5); while the herbivorous, being destitute of that appendage, have the opening of the shell entire (Tab. 23, Fig. 3, 5, 6, 7). Some tribes are exclusively marine, others live only in freshwater, while many are restricted to the brackish waters of estuaries. Their geographical distribution is alike various : certain forms (the *Cephalopoda*) inhabit deep waters only, and are provided with an apparatus by which they can rise to the surface; while others are littoral, that is, live only in the shallows along the sea shores ; many exist in quiet, others in turbulent waters ; some are gregarious, like the oyster, while others live in small groups. All these varieties of condition are more or less strongly impressed on their shelly coverings, which may be considered as their external skeletons,* and the experienced conchologist is enabled at once, by the peculiar characters of the shell, to determine the economy and habits of the animal, and consequently the physical condition in which it was placed. In this point of view, fossil shells become objects of the highest importance to the geologist, since they are frequently the only records of the early condition of our planet. But I must return from this

* See an interesting paper on Shells, by Mr. Gray, of the British Museum. Philosophical Transactions.

digression, and review the phenomena presented by
the several groups of the tertiary formations.

11. MINERALOGICAL CHARACTERS OF THE
TERTIARY SYSTEM.—The predominating characters
of the tertiary system, as I have already mentioned,
are alternations of marine strata with those of
lacustrine and freshwater origin. A large proportion are arenaceous, with intervening beds of clays
and marls. Shingles, the remains of ancient sea-
beaches, abound in some localities, and form a
conglomerate or puddingstone, as that of Hertfordshire (page 76); or a ferruginous breccia, as at
Castle Hill, near Newhaven, on the Sussex coast.
The ruins of the chalk are everywhere recognisable
in the beds of chalk-flints, which contain shells
and zoophytes peculiar to the cretaceous system.
Large boulders are of frequent occurrence, and
may, perhaps, be referred to the newest beds of
the series. In the vicinity of Brighton, blocks of
ferruginous breccia lie near the old church, and on
the race-course; and large masses of quartzose sandstone, of a saccharine structure, are seen at Falmer
and in Stanmer Park: a remarkable rock of this
kind formerly existed in Goldstone Bottom, but is
now destroyed. In many of the gravel beds around
London, are blocks of siliceous breccia, and I have
noticed specimens of considerable magnitude in the
grounds of John Allnutt, Esq. of Clapham Common.
In some of the tertiary formations, limestone pre-
dominates, and alternates with sands and marls of

great variety and brilliancy of colour; with beds of gypsum, and siliceous nodules closely resembling the flints of the chalk. Such are the general features of this system of deposites, which I will now examine more in detail.

The distribution of the tertiary deposites over Europe, appears to be in areas more or less well defined; in our own island, there are the basins of London and Hampshire, and the remains of others in Yorkshire, and in Norfolk and Suffolk. In France, the metropolis is situated within the confines of a tertiary basin; and in the south and north of France, extensive tracts are formed of these deposites; in Auvergne, where they are associated with ancient volcanic eruptions, they constitute a district of unrivalled geological interest. In the Sub-appennines, they are largely developed, and in other parts of Sicily and Italy insensibly pass into vast deposites, which are still in progress of formation.

12. NEWER TERTIARY, OR PLIOCENE DEPOSITES.—From the large proportion of recent species of shells which occur in some of the Pliocene strata, the beds have the appearance of a modern aggregate, as the extensive and beautiful collection from Palermo, before us (for which I am indebted to the kindness of the Marquis of Northampton,) well displays, A low range of hills, rising to an elevation of about 200 feet above the level of the Mediterranean, immediately behind Palermo, is in a great

measure constituted of coarse limestone, formed of
friable shells, which, as you may perceive, are
frequently in an admirable state of preservation;
white and friable in general, but in some examples
preserving their markings and natural polish.   The
elegant and picturesque manner in which they are
occasionally grouped together, as in many of these
specimens, renders them objects of great beauty
and interest.   The shells, with but very few exceptions, belong to species still living in the adjacent seas; a proof that when this limestone was
formed, the same condition of the basin of the
Mediterranean existed as at present, and continued
uninfluenced by the elevation of this portion of
its ancient bed.   In other parts of Sicily, limestone, blue marls, with shelly calcareous breccia,
and gypseous clay, intermingled with volcanic products, occur.   The *Val di Noto* is particularly
mentioned by Mr. Lyell, as presenting a remarkable assemblage of deposites;* and I will quote
his lucid and highly graphic description.   "The
rising grounds of the Val di Noto are separated
from the cone of Etna, and the marine strata on
which it rests, by the plain of Catania, which is
elevated above the level of the sea, and watered by
the Simeto.   The traveller passing from Catania to
Syracuse, by way of Sortina and the valley of
Pentalica, may observe many deep sections of these

* Principles of Geology, vol. iii. p. 388.

modern formations, which rise into hills from one
to two thousand feet in height, entirely composed
of sedimentary strata, with recent shells ; these
are associated with volcanic rocks. The whole
series of strata, exclusively of the volcanic products,
is divisible into three principal groups. 1. The
uppermost, compact limestone in laminated strata,
with recent shells ; total thickness from 700 to
800 feet. 2. Calcareous sandstone, with schistose
limestone. 3. Laminated marls and blue clays."
The whole of the above groups contain shells and
zoophytes of the same species as those from Palermo
which I have just noticed. The large scallop or
Pecten, which you perceive is here so beautifully
preserved, and at the present day is profusely
strewn on the Sicilian shores, is also abundant in
the compact limestone. Leaves of plants and stems
of reeds, are of common occurrence.

13. CRAG OF NORFOLK AND SUFFOLK.—On the
eastern coasts of Essex, Norfolk, and Suffolk, beds
of sand and gravel, abounding in shells and corals,
which are superposed on the blue clay lying on the
chalk, are distinguished by the name of *Crag*, a
provincial term signifying gravel. My late friend,
Mr. Parkinson, first noticed these deposites, and in
the "Organic Remains of a Former World," *
has figured a shell, which was formerly in much request among collectors, the Essex reversed whelk,

* Vol. iii. pl. 6. fig. 5.

(*Fusus contrarius*), in which the spiral convolutions pass from right to left, instead of in the opposite and ordinary direction. Here are several beautiful examples of this fossil shell, collected by my friend Sir Woodbine Parish; they all have the deep ferruginous colour which so commonly prevails in the fossils of the Crag. The Crag first appears at Walton Nase in Essex, and forms the summits of the cliffs on both sides of Hanwell, from a few feet to thirty or forty in thickness. It extends inland along the Suffolk and Norfolk coast, forming a tract of at least forty miles in length; near Ipswich it is spread over a considerable area, and abounds in shells and other marine exuviæ. The series which I now place before you is from collections made by the late Mrs. E. Cobbold, of Holywell Park, near Ipswich; Sir Woodbine Parish; Samuel Woodward,* and Edward Charlesworth,† Esqrs. whose recent investigations have thrown much light on the zoological characters of these deposites. Mr. Charlesworth divides the Crag into two groups; the lowermost series, or Coralline Crag, which is composed of loose sand, and abounds in corals, sponges, and shells, in so perfect a state as to indicate that they lived and died on the spot where their remains are entombed. This series is upwards of fifty feet in thickness, and

---

* Author of "Outlines of the GEOLOGY of NORFOLK."
† The present Editor of the *Magazine of Natural History.*

rests upon a bed of blue clay which will hereafter be noticed. The uppermost, or *Red Crag*, so called from its deep ferruginous colour, consists of sand with shells, which are generally broken and water-worn; the Norfolk Crag appears to be principally composed of these upper deposites. The fossils of the Crag are extremely numerous, they consist of several hundred species of marine shells, some extinct, but the greater part of species now existing in the German Ocean; corals, sponges, and more than a hundred species of microscopic foraminifera; with teeth and scales of fishes. The collection of *Crag* shells on the table was examined by Mr. Lyell, and M. Deshayes, a distinguished French naturalist, by whom more than half the species were considered to be of extinct forms; and the remainder identical with species which now inhabit the German Ocean.*

14. THE SUB-APPENNINES.—The Appennines, that chain of hills which extends through the Italian peninsula, are flanked both on the side of the Adriatic and the Mediterranean, by the Sub-appennines, a low range composed of tertiary marls, sands, and conglomerates, abounding in marine shells of genera and species, which prove that some of the deposites were contemporaneous with the Crag, and that others are referrible to a more ancient epoch. These beds have resulted from the

* Principles of Geology, vol. iv. p 71.

waste of the secondary rocks which form the Appennines and were dry land before these strata were
deposited.*

15. MIDDLE TERTIARY, OR MIOCENE DEPO-
SITES.—In the classification of Mr. Lyell, the term
Miocene designates those tertiary beds in which
recent species of shells occur, but in a much less
proportion than in the preceding division ; seldom
amounting to one-fifth of the whole. As there
are no good types of this group in Great Britain,
I shall merely observe, that marine and fresh-water
deposites possessing the characters here defined occur near Bourdeaux and Montpellier; in Piedmont,
Styria, Hungary, and other parts of the European
continent; but in many instances cited by Mr. Lyell,
the strata seem to merge into one or other term of
the series. I proceed therefore to the consideration
of the *Eocene*, or those tertiary strata which are
of the highest antiquity, and deposited in basins
or depressions of the chalk, where that formation
constitutes the fundamental rock of the country.
Every step of our progress will now be replete with
the deepest interest, and new and singular forms of
being will appear before us. I shall pass rapidly over
the stratigraphical character of these deposites, that
our attention may be more fully directed to the extraordinary organic remains which they inclose.

* Lyell's Principles of Geology, vol. iv. Brocchi, an eminent Italian naturalist, published many years since a valuable
work on the fossil shells of the Sub-appennines.

16. LOWER TERTIARY, OR EOCENE DEPOSITES.
—I propose, in the first place, to describe the geographical distribution and general characters of a
few principal groups of the older tertiary strata;
secondly, to investigate the nature of the more remarkable fossil animals and plants ; and lastly, to
survey those regions of central France, of the Rhine,
and South America, which have been the scenes of
active volcanoes during the tertiary epoch.

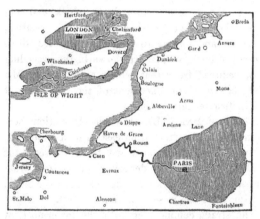

TAB. 19.—TERTIARY BASINS OF PARIS, LONDON, AND HANTS.*

It may be regarded as a singular coincidence, that
the capitals of Great Britain and France are located
on strata of the same geological epoch.   Paris is

* From Mr. Webster's Map in the Geological Transactions.

o

situated on a vast alternation of marine and fresh-
water deposites, lying in a depression of the chalk ;
the latter forming the boundary of the area in
which the city is placed.   London is built on clays,
sands, and shingles, also filling up a basin of the
chalk, which skirts the area of the tertiary strata on
the south, but is open to the sea on the east.   In
Hampshire, a series of contemporaneous deposites,
but with intercalations of lacustrine beds, in like
manner occupy a depression in the chalk, and constitute the basin of the Isle of Wight.   The relation,
situation, and comparative extent of these three
systems of tertiary beds, are shown in the preceding
map, reduced from that which accompanied Mr.
Webster's first announcement of the characters of
the British tertiary formations.

17. THE PARIS BASIN.—The formation here
delineated is in extent from east to west about 100
miles, and 180 from north-east to south-west; the
total thickness of the deposites, or, to use other
terms, the depths passed through to reach the chalk,
varying from one to several hundred feet.

The strata, commencing with the lowermost, or
most ancient, present the following characters :—

1. *The lowermost.* Chalk flints, broken and partially rolled, sometimes conglomerated into ferruginous breccia.   A layer of this kind is very common
on the South Downs, immediately below the turf.

2. *Plastic clay, and sand.* Clay and sand, with
fresh-water shells, drifted wood, lignite, leaves, and

fruits; intercalated with limestone containing marine shells.

3. *Siliceous limestone*, with fresh-water and terrestrial shells and plants; and marine limestone, or *Calcaire grossier;* a coarse compact limestone, passing into calcareous sand, and abounding in marine shells.—These beds often alternate, and are considered by M. Constant Prevost to be contemporaneous formations; the marine strata having been formed in those parts of the basin which were open to the sea; and the siliceous limestone, by mineral waters poured into the bay from the south; the continent being situated then, as now, to the south, and the ocean to the north. Partial deposites of milliolite limestone,* almost entirely composed of microscopic chambered shells, occur in this part of the basin.

4. *Gypseous marls, and limestones,* with bones of animals, and fresh-water shells of fluviatile origin. These are supposed to have been discharged by a river which flowed into the gulf; the gypsum being precipitated from water holding sulphate of lime in solution, in the same manner as travertine or carbonate of lime, of which we have already spoken. (page 48.)

5. *Upper marine formation,* consisting of marls, micaceous and quartzose sand, with beds of sandstone abounding in marine shells.

6. *Upper fresh-water marls,* with interstratified

---

* So called from its inclosing immense quantities of a minute shell, called *Milliolite.*

beds of flint, containing seed-vessels of aquatic plants (*Charæ*), and animal and vegetable remains. These beds are attributed to lakes or marshes, which existed after the marine sands had filled up the basin.

From this rapid sketch, we perceive that the strata which fill up the Paris basin, have been produced by a succession of changes, that readily admits of explanation by the principles so ably enforced by Hutton, Playfair, and Lyell, and explained in the previous lecture. Here we have an ancient gulf of the chalk, which was open to the sea on one side, while on the other it was supplied by rivers charged with the spoils of the country through which they flowed, and carrying down the remains of animals and plants, with land and river shells; and there were occasional introductions of mineral waters. Oscillations of the level of the land and sea took place, and thus admitted of new accumulations upon the older previous deposites; and lastly, the country rose to its present elevation. Changes of this kind, we have already seen, are in actual progress at the present moment, and afford a satisfactory elucidation of these interesting phenomena. I reserve my remarks on the organic remains of the Paris basin to the next section of this discourse, and pass to the examination of the analogous deposites in our own island.

18. THE LONDON BASIN.—The tertiary strata on which the metropolis of England is situated are

spread over a considerable area, which is bounded on
the south by the North Downs; extends on the west
beyond High-elm Hill, in Berkshire; and on the
north-west is flanked by the chalk hills of Wiltshire, Berkshire, Oxfordshire, Buckinghamshire,
and Hertfordshire. On the east it is open to the
sea; the Isle of Sheppey, situated in the mouth of
the Thames, being an outlier of the same deposit.*
It spreads over Essex, a considerable part of Suffolk, Epping and Hainault forests, the whole of
Middlesex, and a portion of Bucks. In the immediate vicinity of the metropolis, a stiff clay of a
bluish-black colour, abounding in marine remains,
constitutes the great mass of the materials which
fill up this ancient gulf of the ocean. Immediately
upon the chalk, however, there occur thick beds
of sand and clay, called Plastic clay (from its analogy to the *Argile Plastique* of the Paris basin),
in which fresh-water shells, plants, and drifted
wood, have been found in some localities. In other
instances, layers of green sand lie upon the chalk,
which at Reading contain immense quantities of
oyster-shells: a similar accumulation of shells has
been observed in Surrey, at Headley, a few miles
from Reigate, by Mr. Peter Martin, jun. of that
town. At Bromley in Kent, there is a bed of
oyster shells with pebbles of chalk-flints, which are
cemented together by a calcareous deposit into a

* See Mr. Webster's paper in the Geological Transactions;
and Conybeare and Phillips' Geology of England and Wales.

remarkable conglomerate, which is used for orna-
mental grottoes. The London clay is found immediately beneath the gravel which so generally forms
the sub-soil of the metropolis ; it is of great extent,
and varies from 300 to 600 feet in thickness. This
clay forms a dark, tough soil, and has occasional
intermixtures of green and ferruginous sand, and
variegated clays. It abounds in spheroidal nodules
of indurated argillaceous limestone, internally filled
by veins of calcareous spar, or sulphate of barytes,
disposed in a radiating manner from the centre
of the nodule to the circumference. From the appearance of partitions which this character confers,
these concretions are commonly known by the
name of Septaria : shells and other organic remains
frequently form the nucleus of these nodules, which
are used in prodigious quantities for cement. The
specimens on the table are from Highgate and
Bognor ; two from the latter locality, presented to
me by my friend Dr. Hall, contain beautiful examples of an extinct species of Nautilus. The
septaria are commonly disposed in horizontal lines,
and lie at unequal distances from each other.
Brilliant sulphuret of iron abounds in the clay,
and is seen in this septarium, as well as in many of
the organic remains. Crystallized sulphate of lime,
or selenite, is also common in these as in other argillaceous strata. The cuttings through Highgate
Hill, to form the archway ; the excavations in the
Regent's Park ; and more recently the tunnels carried

through a part of the same ridge of clay at Prim-
rose Hill, in the line of the Birmingham railroad;
and the explorations, by wells, over the whole area
around London, have brought to light such prodigious quantities of organic remains, that the fossils
of this deposit are almost universally known. The
admirable work of my late friend Mr. Sowerby,
called early attention to these testaceous remains,
the first plate in his Mineral Conchology being
devoted to the " Nautili of the London Básin."
Immense numbers of marine shells of extinct species; crabs, lobsters, and other crustacea; teeth of
sharks, and remains of other genera of fishes; bones
of crocodiles and turtles; leaves, fruits, stems of
plants, and rolled trunks of trees, perforated by boring shells,—occur throughout these deposites, but
are located in greater abundance in some spots than
in others. The clay and gravel pits at Woolwich, on
the banks of the Thames, abound in univalve marine
shells; and at Plumstead, Bexley, and other places,
shells occur in clay, and indurated argillaceous
limestone.

19. THE ISLE OF SHEPPEY.—The Isle of Sheppey
is entirely composed of the London clay, and the
depth of the beds is upwards of 550 feet. It has
long been celebrated for its organic remains; and
I may observe, that the discovery of seed-vessels
and stems of plants in pyritous clay, in a visit which
I made to Queenborough, when a boy, tended to
confirm my early taste for geological researches.

The cliffs on the north of the island, are about
200 feet high, and consist of clay, abounding in
septaria, which are washed out of the cliffs by the
action of the sea, and are collected for cement.
The organic remains are, however, unfortunately
so strongly impregnated with pyrites, that the col-
lector often finds the choicest fossil fruits in his
cabinet, like the fabled apples of the Dead Sea, one
moment perfect and brilliant, and the next decomposed and changed to dust; leaving only an efflo-
rescent sulphate of iron.   The animal and vegetable
remains that occur in the blue clay of the metropolis,
abound in profusion in the Isle of Sheppey.

20. FOSSIL FRUITS. — Seed-vessels, and stems
and branches of trees, of a tropical character, probably drifted by currents into the gulf of the
London basin, are in such abundance and variety,
that the existence of a group of spice islands
seems necessary to account for so vast an accumulation of vegetable productions.   The seed-vessels

TAB. 20.—FOSSIL FRUITS, FROM SHEPPEY.

found at Sheppey are referrible to several hundred
species; some are related to the cardamom, date,

areca, cocoa, (*Cocos Parkinsonis*) : and one berry
bears much resemblance to the coffee. (Tab. 20).
The wood found in the Sheppey clay is generally
of a dark colour, with the ligneous fibres and circles
of growth well developed ; it is often veined with
brilliant pyrites, and the fissures and cavities are
frequently filled with this miner al.It is rarely that
any considerable mass of wood is found free from
the ravages of a species of teredo, resembling the
recent *Teredo navalis,* or *borer,* which inhabits the
seas of the West India islands. The tubular shells
sometimes remain, but their cavities, as well as the
perforations in the wood, are filled with pyrites,
indurated clay, argillaceous limestone, or calcareous
spar ; and specimens, when cut and polished, exhibit interesting sections of the meandering grooves
of the Teredines. In this specimen, which I picked
up on the banks of the canal in the Regent's Park,
the grain of the wood, with the shells, and their excavations, are beautifully displayed.

21. UPPER MARINE, or BAGSHOT SAND.—At
Highgate and Hampstead, Purbright and Frimley
Heaths, in Surrey, and on Bagshot Heath, extensive
beds of sand occur, with but few traces of organic
remains ; those hitherto observed are marine shells.
The boulders and masses of saccharine sandstone,
which are abundant in some of the chalk valleys and
on the flanks of the Downs, are called Sarsden-stone,
or Druid sandstone, from being the principal material
employed in the construction of Stonehenge, and

other Druidical monuments, are supposed to have
been derived from the sand-beds, which overlie the
London clay in the places above named : they may,
however, have belonged to the sands which lie between the clay and chalk. The wastes and unpro-
ductive heaths around the metropolis, are sites of
these arenaceous deposites, which also form the
sub-soil of that charming and picturesque spot,
Hampstead Heath. The gravel and shingle, associated with the sands, have unquestionably been
derived from the ruins of the chalk formation.

22. ARTESIAN WELLS.—As from the alternation
of porous, arenaceous strata, with stiff or impervious
beds of clay, the artificial perennial fountains, called
Artesian wells, are of frequent occurrence in the
vicinity of the metropolis, I will offer a few remarks
on the phenomena of springs. The descent of
moisture from the atmosphere upon the earth,
and its escape into the basin of the ocean, by
the agency of streams and rivers, were noticed in
the first lecture. The rain descending on a
gravelly or porous soil must, of course, descend
through it, till its progress is arrested by a
clayey or impervious stratum, which thus forms
a natural tank or reservoir, collects the water,
and a subterranean pool or canal is produced,
according to the direction and configuration of
the upper surface of the clay. This state of
things will continue, till, by an increased supply,
the waters rise above the level of the basin, or

channel, and overflowing, escape, either through the
porous strata, or by fissures in the solid deposites,
to another level. If the course of the waters be subterranean, the softer portions of the rock are worn
away, and chasms or caverns are formed. Subterranean rivers and streams, of great extent, occur in
many of our mines; but if the water finds its way to
the surface, a spring is said to burst forth. This is
the nature of all springs, except those which arise
from great depths, and are probably dependent on
the condensation of steam, evolved through fissures
by volcanic agency; such are the thermal waters
of many countries. Streams impregnated with
the mineral substances contained in the strata
through which they flow, are called mineral waters.
Those in the tertiary strata near Epsom, contain
sulphate of magnesia, whence the name of Epsom
Salts, given to this substance wherever it occurs.
But it may happen, that strata which are pervious,
alternate with others which are not so; or may form
a basin, the area of which is partially filled with
clay, through which water cannot pass: in such a
case, it is obvious that the bed of sand beneath the
clay, fed by the rain which descends on the un-
covered margin of the basin, must form a reservoir,
and the water gradually accumulate beneath the central plateau of clay, through which it cannot escape.
If this bed of clay be penetrated, either by natural
or artificial means, the water must necessarily rise
to the surface, and may even be thrown up in a jet,

to an altitude which will depend on the level of the
fluid in the subterranean reservoir; such is the phenomenon observable in the Artesian wells around
London. The blue clay confines the water contained
in the sands beneath, and the engineer perforates
this bed, introduces tubes, and taps the natural tank;
by this method, the perennial fountains of Tooting,
Hammersmith, Fulham, &c. have been obtained.*
Of the practical utility of geological knowledge,
even in the common operation of sinking a well, I
once had a striking proof. A gentleman residing
in Sussex, on the borders of the Forest Ridge, who
had seen with admiration the perpetual springs in
the environs of the metropolis, determined to form
one in his grounds at —— Park. Accordingly,
a person conversant with the construction of certain
wells around London was employed, and the necessary apparatus obtained: but the engineer, being
wholly ignorant of the nature of the strata, carried
his operations to a great depth, through the beds of
the Wealden sand, of which the district is composed,
and, of course, without success, as the least geological knowledge of the strata of the country would
have foretold. The undertaking, after considerable
labour and expense, was abandoned.

23. THE HAMPSHIRE, or ISLE OF WIGHT BA-

* Consult Dr. Buckland's Bridgwater Essay, p. 561; and
an admirable Essay on Artesian Wells, in that excellent scientific periodical, the Mining Journal, conducted by H. English,
Esq., F.G.S.

SIN.—The London basin has presented us with but little analogy to the series of alternate marine and freshwater deposites of that of Paris; but in Hampshire and the Isle of Wight, there is an extensive suite of tertiary strata, which, like that of France, is made up of clays, sands, and limestones, containing marine, with intercalations of freshwater remains. This series of tertiary beds extends over a considerable district. On the east, a small outlier of the lower beds appears at Castle Hill, near Newhaven; but proceeding to the westward of Brighton, the London clay rises to the surface beyond Worthing, and forms the tract between the Downs and the sea-shore. The inland boundary stretches by Chichester, Emsworth, and Southampton, to Dorchester; and the clay is spread over the whole area of the New Forest and the Trough of Poole, being flanked by the chalk on the north, north-east, and north-west, and open to the sea on the south. The Isle of Wight, although now separated from the main land, is a disrupted mass of the formations of the south-east of England, as I shall explain in a future lecture. My observations on the present occasion are confined to the tertiary strata; and I will only observe, that the chalk basin has been broken up, and that in some instances, both the chalk and the superimposed sands, clays, and gravel, are thrown into a vertical position. A remarkable and well-known instance of this phenomenon occurs at Alum Bay, so called from the

alum, formerly extracted from the decomposing
pyrites, with which the clay abounds.

TAB. 21.—ALUM BAY, ISLE OF WIGHT.
(Drawn by Miss Susan Chassereau.)

24. ALUM BAY.—This view conveys a general
outline of the bay ; (*a*) represents the vertical
chalk, and (*b b*) the corresponding marine tertiary
strata, consisting of sands and clays of an infinite
variety of colour, and containing abundance of
shells ; advantage is taken of the diversified tints
of the sand to imitate the drawings of landscapes,
&c. in glass vessels, which are sold to visitors.
The appearance of Alum Bay is thus graphically
described by Mr. Webster, whose able memoir
on the strata above the English chalk, formed a
new era in British Geology, and raised our tertiary

series to an importance equal to that of the Paris basin. " The clays and sands of Alum Bay afford one of the most interesting natural sections that can well be imagined. They exhibit the actual state of the strata immediately above the chalk, before any change took place in the position of the latter. For, although the beds of which they are composed are quite vertical, yet, from the nature and variety of their composition, and the regularity and number of their alternations, no one who views them can doubt that they have suffered no change, except that of having been moved with the chalk from a horizontal to a vertical position. The colour of these sands and clays embraces every variety of green, yellow, red, crimson, ferruginous, white, black and brown." Beds of pipe-clay also occur; some of these contain layers of wood-coal, with branches and leaves of vegetables. This coal burns with difficulty, and emits a strong sulphureous smell; masses are constantly drifted by the sea and thrown on the sea-shores near Brighton, where this substance was formerly used as fuel by the poorer inhabitants.*

25. LONDON CLAY OF THE HAMPSHIRE BASIN. —The London clay extends over the greater portion of the area of the Hampshire basin, its peculiar fossils abounding in many localities. Castle Hill, near Newhaven, which has been already mentioned, is a series of sands, marls, and clays, with beds of

See Geology of the South Downs, p. 261.

oyster-shells and of shingle that occupy the upper part of the hill, and rest upon the chalk which forms the lower fifty feet of the cliff. The sub-sulphate of *alumine*,* a mineral substance peculiar to this locality, occurs in the ochraceous clay which is in immediate contact with the chalk. Selenite abounds in the gypseous marls ; a layer of wood-coal, a few inches thick, contains impressions of plants ;† and the argillaceous beds contain marine shells in such abundance, that some of them are mere masses of compressed shells, held together by argillaceous earth. The oysters are consolidated into coarse stone, and where pebbles enter into the composition of this concrete, a close resemblance is presented to the Bromley oyster-breccia. Teeth of sharks and a few fresh-water shells have been collected in these strata. At Chimting Castle, near Seaford, on the eastern escarpment of the valley of the Ouse, olive-green sand, and a ferruginous conglomerate of chalk-flints, lie upon the chalk, thus determining the original extension of the tertiary beds along the Sussex coast.‡ Proceeding westward from Brighton, the London clay is perceived near Worthing, emerging from beneath those newer deposites which, as we have already seen, contain remains of elephants. At Bognor, an arenaceous

* British Mineralogy, Tab. 499.   Geology of the South-East of England, p. 56.
† Fossils of the South Downs, Pl. viii. figs. 1, 2, 3, 4.
‡ Geology of the South-East of England, p. 62.

limestone, full of the usual shells of the *Calcaire grossier* and London clay,* constitutes a group of low rocks, which in another century will have entirely disappeared. The beauty and variety of the shells, particularly of the Nautili, and of the perforated fossil wood, render these organic remains objects of great attraction. The series on the table contains almost every species that has been discovered in these rocks; and Sir Woodbine Parish, Dr. Hall, R. Skynner, Esq. Mr. Parker of Shoreham, and Mr. Baxter of Lewes, have contributed interesting examples to my collection. In the blue clay at Bracklesham Bay, on the western coast of Sussex, and at Stubbington, fossil shells may be obtained at low-water in profusion; and Hordwell Cliff, in Hampshire, has so long been celebrated for similar productions, that its elegant shells are seen in every collection of organic remains. In all these localities the shells are of the same genera and species with those of the contemporaneous deposites of France.

26. FRESH-WATER STRATA OF THE ISLE OF WIGHT.—The great peculiarity and interest of the Isle of Wight basin, as compared with that of London, consists in the existence of strata containing almost exclusively fresh-water shells, and affording in a few instances remains of the animals that occur in the vicinity of Paris. Mr. Webster has arranged the various tertiary beds of

* Fossils of the South Downs, p. 271.

P

Hampshire, in the following sub-divisions:—1st. *lowermost;* Plastic clay and sand. 2d. London clay. 3d. Fresh-water deposites,—sandy calcareous marls, with immense quantities of fresh-water shells. 4th. Clay and marl, abounding in marine shells; very generally of different species from those in the London clay. 5th. Upper fresh-water deposit; yellowish white marl, and calcareous limestone, employed for building ; nearly sixty feet in thickness, and almost one entire mass of fresh-water shells.

The above series is well developed at Headon Hill, where the fresh-water strata, which succeed the marine beds of Alum Bay, lie in a nearly horizontal direction.   At Binstead, near Calbourne, and Morley, quarries 'have been opened in the fresh-water limestone ; and bones of *Anoplotheria,* and of a species of *Moschus,* have been discovered.

27. ORGANIC REMAINS OF THE PARIS, LONDON, AND HANTS BASINS.—I can attempt but a brief description of the fossils of the strata which we have now surveyed, so numerous are the relics of the inhabitants of the ancient land and water which they inclose. To condense my remarks as much as possible, I shall select the fossils of the Paris basin as the types of the zoological characters of the older tertiary epoch, and include notices of such species as occurring in British and other localities, may be requisite for the illustration of the subject.

28. PLANTS.—Fossil wood occurs in vast abundance, particularly in large trunks and branches, which appear to have been drifted, and are full of perforations inclosing shells of boring mollusca. Bognor rocks, the clay around London, Isle of Sheppey, &c. abound in specimens of this kind. The wood is dicotyledonous, that is, like the oak, ash, &c. its mode of increase was by annular circles of growth, as will be explained in the Lecture on Fossil Botany. Leaves and stems of palms have been found in the Paris basin, and in the Isle of Sheppey, &c.; and a trunk of a tree allied to the palm, nearly four feet in diameter, at Soissons. Fruits belonging to trees allied to the areca, pine, fir, cocoa-tree, &c. have been discovered in several localities. Accumulations of vegetable matter, in the state of lignite or *brown coal*, occur at Bovey Tracey in Devonshire, and in various parts of France. Amber, and a substance which has been called Highgate resin, are occasionally imbedded in these deposites.

ZOOPHYTES.— Polyparia occur in some parts of the marine strata, but they are not numerous; several species of turbinolia, caryophillia, fungia, and other corals, are figured and described by authors. I have a few specimens from Grignon : the modern tertiary, (those of Palermo,) abound in flustra and spongia.

29. SHELLS OF THE TERTIARY STRATA.—So numerous are the shells of the tertiary epoch, already determined by naturalists, that they exceed one-half

of the known living species, amounting to nearly three thousand.   We have already seen that some of the strata are almost entirely composed of these remains

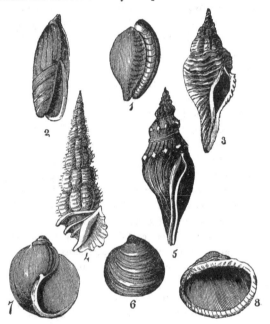

TAB. 22.—MARINE SHELLS OF THE PARIS BASIN.

Fig. 1. *Cyprea inflata.*   2. *Ancilla canalifera.*   3. *Fusus uniplicatus.*
4. *Cerithium lamellosum.*   5. *Pleurotoma dentata.*   6. *Lucina sulcata.*
7. *Ampullaria sigaretina.*   8. *Pectunculus angusti-costatus.*

in a broken and compressed state: at Newhaven, many seams in the argillaceous beds wholly consist of shell-dust.   In other localities the shells occur in the most perfect state ; and Grignon, a few leagues from Paris, has long been celebrated for its beautiful productions of this kind, many hundred species of shells peculiar to the older tertiary strata having been collected in one spot alone : these shells belong to the *Calcaire grossier*, and many of the species occur in the London and Hampshire basins, and Bognor rocks. I have selected a few specimens, from the many hundreds in my possession, to convey an idea of their usual characters and appearance. (Tab. 22.)

Although, in mentioning the names of these shells, I do not expect that any but the scientific inquirer will endeavour to fix them on the memory, yet it is useful to point out to you the characters which prevail in these tertiary beds ; for, as I have already stated, certain fossils are peculiar to certain strata, and the experienced geologist can often, at a glance, determine the relative antiquity of a deposit by an examination of a few species of shells. The whole of these forms must be familiar to you, because they belong to genera which have species that still live in our present seas. The *Cyprea*, or cowry, fig. 1, and the *Ancilla*, or olive, fig. 2, are well-known types. The *Cerithium*, fig. 4, belongs to a genus which is most abundant in the sands of the Paris basin ; and is remarkable for the elegance, number, and variety of the species, which exceed by three times that of their living analogues. The *Cerithium giganteum* attains a considerable magni-

tude.   Some masses of the Bognor rock are almost
wholly composed of a species of *Pectunculus.*   The
Ampullaria, (Tab. 22, fig. 7,) abounds at Grignon,
and is commonly in a beautiful state of freshness.
You will recollect that the channel in the aperture
of the shells (Tab. 22, figs. 2, 3, 4, 5,) indicates that
the animals belonged to the carnivorous tribes.

TAB. 23.—FRESH-WATER SHELLS OF THE PARIS BASIN.

Figs. 1, 2. *Bulimus conicus.*   3, 4. *Cyclostoma mumia.*   5. *Lymnea
effilea.*   6, 7, 8. *Planorbis.*

TERTIARY FRESH-WATER SHELLS.—It has already been observed, that the shells of the mollusca
which inhabit fresh water, possess characters by
which they may be readily distinguished from the

marine species. This small selection from the fresh-
water beds of Paris will serve to elucidate my
observations.

The general appearance of these shells will bring
to your recollection the species which inhabit our
ponds and rivers; particularly the large thin snail,
fig. 5, and the discoidal shells, 6, 7, 8; while figs.
3 and 4 resemble a shell often found on the banks
of lakes. At Headon Hill, and at Binstead in the
Isle of Wight, the clay and limestone are filled with
the remains of several species of Planorbis and
Lymnea.

30. NAUTILI, AND OTHER CEPHALOPODA.—
There are several species of Nautilus which abound
in the tertiary strata ; those inclosed in the
septaria, or indurated argillaceous nodules, of the
London clay at Highgate, Sheppey, and Bognor,
possess considerable beauty, and admit of being cut
in sections, which admirably display the internal
structure of the shell. I shall, however, defer an
explanation of their mechanism to the subsequent
lecture, when analogous fossil genera will come
under our notice. My observations will now be
restricted to an interesting division of the *Cephalopoda* (as those mollusca are termed whose head
is surrounded by the organs of motion, or feet),
called *Foraminifera*, which comprehends many genera, and several hundred species, the greater part
being microscopic, and analogous to the recent
forms which inhabit the Mediterranean. These

bodies are entirely different from the testaceous
habitations of snails, periwinkles, &c. : they are,
in truth, not an external but an internal apparatus ;
and it is supposed, that, in addition to their serving
as a point of attachment and support to the soft
body of the animal, they acted as a buoy, which
could be made heavier or lighter at pleasure, and
by which the animal was enabled either to sink
or swim.   The fossil called *Nummulite* (from its
resemblance to a coin) affords a familiar example

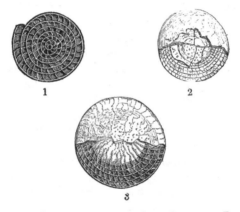

1   2

3

TAB. 24.—NUMMULITES, FROM THE GREAT PYRAMID OF EGYPT.

(Collected by Dr. George Hall, of Brighton.)

Fig. 1. *Transverse Section of a Nummulite.   Figs.* 2, 3. *Nummulites, with
the external plate partially removed.*

of the structure of these bodies.   The nummulite
is of a lenticular, discoidal form, and varies in size

from a mere point to an inch and a half in diameter.
The outer surface is generally smooth, and marked
with fine undulating lines. On splitting the shell
transversely, it is found to be composed of several
coils, which are divided into a great many cells
or chambers by oblique partitions (Tab. 24, fig. 1),
apparently having no communication with each
other, but which, it is probable, the animal had
the power of filling with fluid, or air, through
foramina or pores; whence the name of the order.
To my valued friend, Dr. Hall (physician to the
Sussex Hospital), I am indebted for the specimens
exhibiting this structure, which I now place before
you. They are from the limestone formed of nummulites (Tab. 24,) held together by calcareous
cement, which constitutes the foundation rock of
the Great Pyramid of Egypt, and of which it is
in part constructed. Strabo alludes to the nummulites of the pyramids, under the supposition that
they are lentils which had been scattered about
by the workmen, and become converted into stone.
This polished, silicious pebble, presented me by
Lord Northampton, is also from Egypt; the markings on the surface are sections of the inclosed
shells. The nummulite is one of the most widely
diffused of fossil shells, its remains forming whole
chains of calcareous hills: it is not confined to the
tertiary, but occurs also in the secondary formations, constituting immense beds in the Alps
and Pyrenees. The blue clay at Bracklesham

and at Stubbington, and the calcareous sandstone of Emsworth and Bognor, in Sussex, abound in nummulites. In North America limestone occurs near Suggsville, constituting a range of hills about 300 feet in height, which is entirely composed of one species of nummulite. The limestone is porous, and contains spheroidal cavities formed by the decomposition of the organic remains.*

Tab. 25.—Fossil Crab, from Malta.
(Cancer Macrochelus.)

31. Crustacea and Fishes.—Crabs and lobsters, of species related to the recent, several of which are described by MM. Brongniart and Desmarest, in their beautiful work on Fossil Crustacea, have been found in the clay of Highgate, and in the Isle of Sheppey. The external configuration of the shell or crustaceous covering of these animals,

* Dr. Martin's Synopsis of the Organic Remains of the Cretaceous Group of North America. 8vo. Philadelphia.

being in conformity to the viscera which they inclose and are intended to protect, the naturalist is
able, by an accurate acquaintance with the characters of the living species, to point out the relation or
difference of the fossil, even though the carapace or
buckler alone remains; and the size and situation
of the heart, stomach, &c. may thus be readily determined.   This remarkably fine crab (Tab. 25,) is
from Malta; it shows the state of perfection in
which fossil crustacea are sometimes discovered.

   In the cream-coloured limestone of Pappenheim,
crustacea, allied to the shrimp, lobster, and cray-fish,

TAB. 26.—FOSSIL PRAWN, FROM PAPPENHEIM.

(*Palæmon Spinipes.*)

are often met with ; and many beautiful specimens
are figured by foreign authors.   Knorr's splendid

work, "Monumens des Catastrophes que le Globe
terrestre a essuiées," in particular, has coloured representations of crabs, astacidæ, &c. : the specimen
here figured (Tab. 26,) exhibits the extraordinary
state of preservation of these remains.

In the Paris basin alone, there have been discovered seven or eight species of fishes, of extinct
genera. The teeth of several kinds of sharks
(*Lamna*) occur every where, and are known by the
name of "Birds' bills." In the clay of Sheppey
and London, beautiful fossil fishes have been found,
the scales possessing a metallic lustre, from an impregnation of sulphuret of iron. But I must pass
cursorily over these remains, as well as those of
crocodiles, turtles, and tortoises, which are imbedded in these deposits, or I shall far exceed my
limits.

32. FOSSIL BIRDS.—In the gypseous building-
stone of Montmartre, M. Cuvier found many bones
which possessed characters peculiar to the skeletons
of birds; and after much research he was enabled
to determine several fossil species, belonging to
the genera Pelican, Sea-lark, Curlew, Woodcock,
Buzzard, Owl, and Quail. In some examples there
are indications of the feathers, and even of the air-
tubes. Sometimes the skeleton is wanting, but a
pellicle of a dark brown substance points out the
configuration of the original. (Tab. 27.)

Not only are the skeletons and feathers of birds
found in the tertiary strata, but even the eggs of

aquatic birds occur in the lacustrine limestone
of Auvergne. We have already noticed that eggs

TAB. 27.—FOSSIL BIRD, FROM MONTMARTRE.

of turtles are daily in the course of fossilization on
the shores of the Isle of Ascension.

33. CUVIERIAN PACHYDERMATA. — We have
next to consider the fossil remains of the Mammalia
whose skeletons were entombed in the mud of the
waters which formerly occupied the area of Paris,
and which the genius of Cuvier has again, as it were,
called into existence. The forms of these extinct
creatures are now as familiar to us as our domestic
animals, and even the names of Palæotheria and

Anoplotheria are almost become household words. The gypsum quarries which are spread over the flanks of Montmartre have long been known to afford fossil bones; but, although specimens occasionally attracted the notice of the naturalists of Paris, and collections were formed, yet no one appears to have suspected the mine of wonders which the rocks contained, till the curiosity of Cuvier was awakened by the inspection of a large collection of these bones, after he had successfully applied the laws of comparative anatomy to the investigation of the fossil Elephants and Mammoths. He had previously paid little or no attention to the partial accounts of fossil bones found in the vicinity of Paris, although in 1768 Guettard had figured and described many bones and teeth. Cuvier now, however, perceived that a new world was open to his researches, and he soon, by zeal and energy, obtained an extensive collection, and found himself (to use his own expression) in a charnel-house, surrounded by a confused mass of broken skeletons of a great variety of animals. To arrange each fragment in its proper place, and to restore order to these heaps of ruins, seemed at first a hopeless task; but a knowledge of the immutable laws by which the organization of animal existence is governed, soon enabled him to assign to each bone, and even fragment of bone, its proper place in the skeleton; and the forms of beings hitherto unseen by mortal eye arose before him. " I cannot," says this illustrious philosopher, in all

the enthusiasm of successful genius, " express my
delight on finding how the application of one
principle was instantly followed by the most trium-
phant results. The essential character of a tooth,
and its relation to the skull, being determined, immediately all the other elements of the fabric fell
into their places; and the vertebræ, ribs, bones of
the legs, thighs, and feet, seemed to arrange themselves even without my bidding, and precisely in
the manner which I had predicted." The principles of comparative anatomy enumerated in the
second lecture will have prepared you for this result; and I now need not dwell on the application
of the laws of co-relation of structure by which the
animals of the Paris basin have been restored.

TAB. 28.—ANIMALS OF THE TERTIARY EPOCH.

*Fig.*1. *Anoplotherium gracile.* 2. *Palæotherium magnum.* 3. *P. minus.*
4. *Anoplotherium commune.*

This group of figures, from Cuvier's restorations,
is indeed a splendid triumph of Palæontology.

The examination of the fossil teeth at once showed that the animals were herbivorous, the enamel and ivory being disposed in the manner already explained (page 119); and the crown of the tooth composed of two or three simple crescents, as in certain pachydermata; thus differing from the ruminants, which have double crescents, and each four lines of enamel. Following out the inquiry, Cuvier at length established that the great proportion of bones and teeth belonged to two extinct genera of pachydermata, bearing an affinity to the Tapir, Rhinoceros, and Hippopotamus. Almost every one is familiar with the form and habits of the two last animals; but the Tapir is not so well known. Of this genus there are several living species, all natives of tropical climes. The Malay Tapir (a stuffed specimen of which is placed on the lobby of the British Museum,) sometimes attains eight feet in length, and six in circumference. It has a flexible proboscis, a few inches long; its general appearance is heavy and massive, resembling that of the Hog. The eyes are small, the ears roundish; the skin is thick and firm, and covered with stout hair; the tail short. It inhabits the banks of lakes and rivers, and has been observed to walk under water, but never to swim.*

34. PALÆOTHERIA, AND ANOPLOTHERIA. — The fossil animals are divided into the genera

* Griffiths' Animal Kingdom, vol. iii. p. 434.

Palæotherium (*ancient wild beast*); and Anoplotherium (*unarmed wild beast*), so named from the
absence of canine teeth.   I will now describe the
species here figured.

PALÆOTHERIUM MAGNUM, fig. 2.   This animal
was of the size of a horse, but more thick and
clumsy; it had a massive head, and short extremities.   It was like a large tapir, but with differences
in the teeth, and a toe less on the fore feet.   It must
have been from four to five feet in height, which
is about equal to that of the rhinoceros of Java.
From the conformation of the nasal bones, no
doubt can exist of its having been furnished with
a short proboscis, or trunk.

PALÆOTHERIUM MINUS, fig. 3.   Of the size of
the roebuck.   This creature had light and slender
limbs, with the general configuration of the tapir.

ANOPLOTHERIUM GRACILE, fig. 1.   This animal,
to which Cuvier gave the name of *gracile*, from
its elegant proportions, was of the size and form
of a gazelle, and must have lived after the manner
of the deer and antelopes.

ANOPLOTHERIUM COMMUNE, fig. 4; was of the
height of a wild boar, but of a more elongated
form, and had a long and thick tail like a kangaroo,
the feet having two large toes like the ruminants.
It seems probable that it could swim with facility,
and frequented the lakes, in the beds of which its
bones were deposited.

More than fifty extinct mammalia have been

discovered in the older tertiary, and their characters determined by Baron Cuvier. Some are related to the animals we have just described ; as the
Anthracotherium, (so named from the discovery of
its remains in the Anthracite, or lignite of Cadi-
bona,) which held an intermediate place between
the hog and hippopotamus. Six or seven species
of carnivora, an opossum, a squirrel, dormouse, &c.
have also been found in the Paris basin.

In the miocene strata of Touraine and of Darm-
stadt, there is an intermixture of the remains of the
above extinct mammalia with those of the Mastodon, and of genera which still exist. Mr. Murchison
has discovered in Bavaria, bones of the Palæotherium,
Anoplotherium, Anthracotherium, Mastodon, Rhinoceros, Hippopotamus, Ox, Horse, Bear, &c. in
lacustrine deposites, associated with fresh-water
and land shells.

35. FOSSIL QUADRUMANA OR MONKEYS.—At
Sausan, in the department of Gers, are tertiary
deposites, abounding in remains of the rhinoceros, horse, palæotherium, anoplotherium, and other
mammalia. M. Lastel has discovered a jaw of a
Monkey, which, from its proportions, must have belonged to an animal about three feet in height. The
molar teeth in the specimen are worn, and very
closely resemble those of a man of middle age,
reduced to half their natural size. Another fossil
monkey has been found in the hills near the Sutlej,
associated with remains of mastodons, elephants,

crocodiles, turtles, &c. The specimen is the right half of the upper jaw, to which a portion of the orbit of the eye remains attached, and this alone is sufficient to enable an anatomist to determine the nature of the original, the orbits of the Quadrumana being peculiar. Without entering upon details uninteresting to the general inquirer, it may be stated that evidence is thus afforded of the existence of a gigantic species of monkey, contemporaneously with the pachydermata, whose fossil remains occur in the Sub-Himalayas. The important fact is therefore now established, that animals of that type of organization which most nearly resembles the human, existed in the ancient tertiary epochs.*

* THE QUADRUMANA.—These animals come nearest to man in the form and proportion of their skeleton, and of their separate bones ; in the general disposition of their muscular system, and its adaptation for a semi-erect position of the body ; in their great cerebral organization, the perfection and equable development of their senses; their intellectual capacity, and complicated instincts. These most elevated of all inferior animals are fitted to select, obtain, and digest the succulent ripe fruits of trees, and are destined to inhabit the rich and shady forests of tropical climates. They leave to the squirrels and the sloths the buds and leaves ; to the ponderous elephant and rhinoceros the branches and the stems ; and to the beavers, and other rodentia, the dry bark of the trees. Their delicate organization is adapted only for the richest products of the vegetable kingdom ; and the soft and nutritious quality of their food is suitable to the broad enamelled crowns of their molar teeth, which are studded with rounded tubercles : their stomach is simple. With a high cerebral and muscular development, corresponding with their elevated rank in the scale of beings and the

Q 2

36. TERTIARY STRATA OF AIX, IN PROVENCE. —A group of tertiary strata, remarkable as well for their mineralogical character as for the organic remains which they contain, occurs near Aix, a town in Provence, which is situated upon a thick deposit of tertiary conglomerate. The series on the northern side of the valley consists of — 1. Tertiary breccia, (see Pl. V. fig. 3,) lying unconformably on the secondary rocks of oolite, and green sand, which are nearly vertical. 2. Marl, with fishes and insects. 3. Gypsum and gypseous marls, with fishes and insects; leaves of palms, and other plants; fresh-water univalve and bivalve shells, particularly a species of *Cyclas* in great abuudance. The *cyclas* inhabits lakes and marshes, and therefore positively denotes the lacustrine character of the deposites. 4. Fresh-water limestone. To the south, extending towards Toulon, are lacustrine strata of red marl, with compact limestone inclosing shells, gyrogonites, &c. Still farther to the south, beds of grey fresh-water limestone appear; and at Fuveau, (see the section, Pl. V. fig.3,) a series of blue limestone, shale, and coal, which is extensively worked. Fresh-water shells, seed-vessels of Charæ, and other vegetable remains, occur in abundance in the coal-beds and

position of their food, they are the most agile and sportive of all Mammalia; and they are provided with prehensile organs at every point; their teeth, tail, feet, and hands assist in their agile movements, and in their boundings from branch to branch, and from tree to tree. — *Dr. Grant's Lectures on Comparative Anatomy.*

intermediate layers of shale. The section employed to illustrate this description is copied from a rough sketch made by Mr. Lyell on the spot, when he first visited this interesting locality ; and which I greatly value on that account. (The term *Lias* in the lithograph, Pl. V. fig. 3, is an error; the beds of limestone thus denominated are fresh-water tertiary deposites.)

The marls, as you may perceive, in this extensive suite of specimens, collected by my friends Messrs. Lyell and Murchison, (whose admirable Memoir* on these strata is of high interest,) are very finely laminated, and contain insects and fishes in a remarkable state of preservation. The fishes are very numerous ; one small species in particular (*Smerdis minutus*) is found grouped together in every variety of form and position.

37. FOSSIL INSECTS.—But the insects are the most extraordinary fossil remains, appearing as fresh as if enveloped but yesterday. This beautiful example, presented to me by that highly-gifted lady Mrs. Murchison, shows the exquisite preservation of these delicate objects. A few of the most interesting forms are here delineated on a slightly enlarged scale, from the plate accompanying the paper to which I have referred.†

* On the Fresh-water Formation of Aix, in Provence, by C. Lyell, and R. I. Murchison, Esqrs. Edinburgh Philos. Journal, 1829.

† Jameson's Edinburgh Journal, for 1829. Pl. VI,

Mr. Lyell states, that all the insects are referrible to existing genera, and that only one species is aquatic. The anterior *tarsi** are generally obscure,

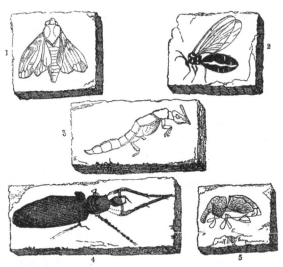

TAB. 29.—FOSSIL INSECTS FROM AIX, IN PROVENCE.

Fig. 1. *Tettigonia spumaria.* 2. *Mycetophila, imbedded while in the act of walking; the articulations of the body distended by pressure.* 3. *Lathrobium.* 4. *Allied to Penthetria holosericea. The hinder legs are broken off, and one of them is reversed, so that the* tarsi *nearly touch the thigh; the* palpi *are long and perfect; the* antennæ *remarkably distinct.* 5. *Liparus, resembling L. punctatus.*

or distorted; but in some specimens the claws are visible, and the sculpture, and even a degree of local colouring are preserved. The nerves of the

* Principles of Geology, vol. iii. p. 211,

wings in the *diptera*, and even the pubescence on
the head, are distinctly seen. Several of the beetles
have the wings extended beyond the elytra, as if
they had fallen into the water while on the wing, and
had made an effort to escape by flight. M. Marcel
de Serres has enumerated nearly seventy genera of
insects, and a few Arachnides, or spiders. The
most curious fact is, that *some of the insects are
identical with species which now inhabit Provence.*
It seems probable that these insects were brought
together from different localities by floods, and
mountain streams; yet, as Mr. Curtis observes, all
of them might have inhabited moist and shady
forests. The laminated marls contain also the
coverings of a fresh-water crustacea, called *Cypris,*
which swarms in our pools and stagnant waters,
and must be familiar to all who have seen the
exhibition of the oxy-hydrogen microscope; living
specimens being commonly shown, and appearing
somewhat like the head and feet of a flea protruding
from an oval case or shield, and swimming by means
of their fine *cilia,* which resemble pencils of hair.
These crustacea shed their cases, some of which are
siliceous and others calcareous, annually, and the
surface of the mud spread over the bottoms of lakes
is often strewed with their relics. The marls of
Aix, as well as of many other fresh-water formations,
abound in fossil *Cyprides,* which sometimes constitute entire seams or laminæ, that alternate with the
marl. The seed-vessels of the *Chara,* a common

plant in our ditches and ponds, also occur in profusion; they were formerly supposed to be shells, and from their peculiar structure received the name of *Gyrogonites*, which they still bear, although their real nature has long been ascertained.* In conclusion, Mr. Lyell observes, " that this series of tertiary deposites differs essentially from those of the London and Paris basins. The great development of regular beds of blue limestone and shale, the quality and appearance of the coal, the thickness of the compact grey, brown, and black argillaceous limestones and sandstones, give to these deposites the aspect of the most ancient of our secondary rocks; and it is only by the various peculiar species of fluviatile and lacustrine shells, the seed-vessels of the Chara, &c. that the comparatively recent date of the whole group is demonstrated."

38. LACUSTRINE FORMATION OF ŒNINGEN.—
Among the tertiary lacustrine formations on the continent, there is one so much celebrated for its organic remains, that I will offer a few remarks on its peculiar characters. Œningen, near Constance, has for centuries been known to contain fossil remains of great beauty and interest. A short, but graphic, memoir by Mr. Murchison,† presents in a few pages the history of this ancient lake. The Rhine, in

* See an Essay on the Fresh-water Marls of Scotland, by Mr. Lyell.

† On a Fossil Fox found at Œningen, by R. I. Murchison, Esq., Pres. G.S. &c. Geological Transactions. 1832.

its course from Constance to Schaffhausen, flows
through a depression of the tertiary marine formation, known by the name of *Molasse*, which forms
hills on both sides of the river, of from 700 to
900 feet in height. In a basin of this molasse, is
a series of strata composed of *marls*, and cream-
coloured, fine-grained, fetid limestone, with laminated white marl-stones, forming a total thickness
of thirty or forty feet. In the marl-stone, leaves
and stems of plants, insects, shells, crustacea, fishes,
turtles, a large aquatic salamander, birds, and a
perfect skeleton of a fox, have been discovered.
The fox was obtained by Mr. Murchison, for whom
I developed it, and removed the stone so as to expose
the entire skeleton : this extraordinary fossil, which
does not differ in its osteology from the recent
species, is figured and described in the Geological
Transactions for 1832. A tortoise, three feet in
length, with the head, neck, tail, and three of the
paws, well preserved, has since been discovered.
Mr. Murchison concludes that these fresh-water
deposites are the contents of a lake, belonging to
the newer pliocene epoch; yet the period of their
formation must have long preceded the present
condition of the country, the Rhine having subsequently worn a channel through them to the depth
of several hundred feet.

39. FOSSIL FISHES OF MONTE BOLCA.—I will
here notice another interesting assemblage of tertiary
strata—the celebrated ichthyolite quarries of Monte

Bolca—and then proceed to the consideration of the
effects of volcanic action during the geological epochs
embraced in this discourse.   Monte Bolca is situated
on the borders of the Veronese territory, about fifty
miles NN.W. of the lagunes of Venice, and forms
part of a range of hills of moderate elevation;
volcanic deposites abound in the neighbouring Vi-
centin; and the summit of the hill is capped with
basalt.*   It is principally composed of argillaceous
and calcareous strata, with beds of a cream-coloured
fissile limestone, which readily separates into laminæ of moderate thickness, and abounds in fishes
in the most beautiful state of preservation.   They
are all compressed flat, but the scales, bones, and
fins remain; their colour is a deep brown, thus
admirably contrasting with the limestone in which
they are imbedded.   Several hundred distinct species are supposed to be contained in these quarries,
and thousands of specimens have been collected;
according to M. Agassiz, all the species, though
related to the recent, are extinct.   From the immense quantities which occur in so limited an area,
it seems probable that the limestone in which they
are imbedded, was erupted into the ocean in a fluid
state by volcanic agency; and that the fishes were
thus suffocated, and surrounded by the calcareous
mass.   Nor is this hypothesis without support,
for on the appearance of a volcanic island in the

* Organic Remains of a Former World, vol. iii. p. 247.

Mediterranean, a few years since, hundreds of dead fishes were seen putrid and floating in the waters; and it cannot be doubted that shoals of fishes might at the same time have been enveloped in the volcanic matter at the bottom of the sea, and become compressed and preserved; when the mud which envelopes them is consolidated, and the bed of the Mediterranean elevated above the waters, these fishes may resemble the ichthyolites of Monte Bolca.

40. TERTIARY VOLCANOES.—In the former lecture I have briefly alluded to volcanic action as still existing, and as having taken place in more ancient periods; and we have abundant proof that during the immense lapse of time comprehended between the earliest and the latest of the tertiary formations, the internal fires of our globe were not dormant. I have already had occasion to observe, how rarely the former geographical relations of a country are preserved, and that although we may be able to pronounce with certainty that this spot was once dry land, — that there flowed a river,—that here is the bed of an ancient sea—yet seldom can we ascribe limits to the one, or trace the boundaries of the other. But there is one remarkable exception—a district, where the most important and striking geological mutations have taken place, and yet the area of these changes still preserves its ancient physical geography — that district is Auvergne, a province in central France.

Nearly a century since, two French academicians, MM. Guettard and Malesherbes, on their return from an exploration of Vesuvius, arrived at Montelimart, a small town on the left bank of the Rhone, where Faujas St. Fond, a distinguished naturalist, was sojourning. These savans were struck with the pavements of the streets, which were formed of short joints of basaltic columns, placed perpendicularly in the ground; and upon inquiry they found, that the stones were brought from the neighbouring mountains of the Vivarais. This information induced them to survey the country; and upon arriving at Clermont, a town with about 30,000 inhabitants, the capital of Auvergne, they were satisfied that the whole region was of volcanic origin; for in the vicinity of the town they discovered currents of lava, black and rugged as those of Italy, descending uninterruptedly from some conical hills of scoriæ, which still preserved the form of craters! "To those who now visit central France, and see on all sides the most unequivocal marks of volcanic agency—the numerous hills formed entirely of loose cinders, and porous and diversified as if just thrown from a furnace, surrounded by plains of black rugged lava, on which even the lichen almost refuses to vegetate,—it appears scarcely credible, that previous to the last half century, no one had thought of attributing these marks of desolation to the only powers in nature capable of producing them. This, however, is perfectly natural, and not without examples. The inhabitants of Herculaneum and Pompeii built their houses with the lava of Vesuvius, ploughed up its scoriæ and ashes, and ascended its crater, without dreaming of their neighbourhood to a volcano which was to give the first proof of its energies by burying them beneath its eruptions. The Catanians regarded as a fable all mention of the former activity of Etna, when, in 1669, half their town was overwhelmed by its lava currents."*

The country which is the site of the extinct volcanoes to which I am about to call your attention, may be described as a vast plain, situated in the department of the Limagne d'Auvergne : it is so remarkable for its fertility, that it is called the Garden of France; a quality attributable to the detritus of volcanic rocks, which enters into the composition of the soil. It is inclosed on the east and west by two parallel ranges of gneiss and granite. Its average breadth is twenty miles, its length between forty and fifty, and its altitude about 1,200 feet above the level of the sea. The surface of this plain is formed of alluvial deposites, composed of granitic and basaltic pebbles, and boulders,

* Geology of Central France, by G. Poulett Scrope, Esq. F.R.S., 1827. Mr. Bakewell was the first English geologist who directed attention to this remarkable district; (Travels in the Tarentaise, by Robert Bakewell, Esq. 2 vols. 8vo. 1823.); subsequently Dr. Daubeny, Messrs. Scrope, Lyell, and Murchison, have severally published highly interesting treatises on the subject.

reposing on a substratum of limestone. Hills, of
various elevations, composed of calcareous rocks,
are scattered over the plain; and the river Allier
flows through the district, over beds of limestone,
except where it has excavated a channel to the
foundation-rock of granite ; or over siliceous sandstone, which also, in many places, prevails. The
hills, composed of calcareous and alluvial deposites,
are relics of a series of beds, which once constituted an ancient plain, at a higher elevation than
the present. Many are surmounted by a crest or
capping of basalt, to which their preservation is
probably attributable; others have escaped destruction from being protected by horizontal layers of
a durable limestone, which I shall presently describe. We have, then, as the ground plan of the
district, an extensive plain, chequered with low hills
of fresh-water limestone, which are capped with compact lava (Plate IV. figs. III. IV.) ; the boundaries
of the plain being formed of ranges of primary
rocks, 3,000 feet in altitude. To the westward the
limestone disappears, and a plateau of granite rises
to a height of about 1,600 feet above the valley of
Clermont, being 3,000 feet above the level of the sea.
This supports a chain of volcanic cones and dome-
shaped mountains, about seventy in number, varying
in altitude from 500 to 1,000 feet from above their
bases, and forming an irregular range nearly twenty
miles in length, and two in breadth. The highest
point of this range is the Puy de Dome, which is

above 4,000 feet above the level of the sea (Pl. IV.
fig. II. 4), and is composed entirely of volcanic
matter; it possesses a regular crater, 300 feet deep,
and nearly 1,000 feet in circumference. Many of
these cones retain the form of well-defined craters,
and their lava currents may be traced as readily as
those of Vesuvius.

41. CRATER OF PUY DE COME.—One of the
most remarkable cones is the Puy de Come, which
rises from the plain to the height of 900 feet;
its sides are covered with trees, and its summits
present two distinct craters, one of which is 250
feet in depth. A stream of lava may be seen to have
issued out from the base of the hill, which at a short
distance, from having been obstructed by a mass of
granite, has separated into two branches; these may
be traced along the granitic platforms, and down the
side of a hill into an adjacent valley, where they
have dispossessed a river of its bed, and constrained
it to work out a fresh channel between the lava and
the granite of the opposite bank. Another cone is
described by Mr. Scrope as rising to the height of
1,000 feet above the plain, having a crater nearly
600 feet in vertical depth, and a lava current, which
first falls down a steep declivity, and then rolls over
the plain in hilly waves of black and scorified rocks.
In one part of this volcanic group, is a circular
system of cones, apparently the produce of several
rapidly succeeding eruptions. " The extraordinary
character of this scene impresses it for ever on the

memory; for there is, perhaps, no spot, even among
the Phlegræan fields of Italy, which more strikingly
displays the characters of volcanic desolation: for
although the cones are partially covered with wood
and herbage, yet the sides of many are still naked;
and the interior of their broken craters rugged,
black, and scorified, as well as the rocky floods of
lava with which they have loaded the plain, have a
freshness of aspect, such as the products of fire
alone could have so long preserved, and offer a
striking picture of the operation of this element
in all its most terrible energy."* A description
of the accompanying sketches† will serve to illustrate these remarks. (Plate IV. fig. III.) The
environs of Clermont.—The town is seen on the
plain or basin, which has been excavated by diluvial agency, since the deposition of the surrounding hills. In front is a basaltic peak (*coloured
green*), crowned by the Castle of Montrognon;
and beyond are basaltic platforms (*green*), on hills
of limestone. In the distance is the primitive
escarpment, forming part of the boundary of the
volcanic district.

Pl. IV. fig. II. Part of the southern volcanic chain
of Puys, exhibiting the broken craters of Chaumont;
from the bases of several, lava currents (*green*)
are seen to issue. No. 1, Montchal; 2, Puys de

---

* Scrope's Geology of Central France.   † The delineations
are reduced sketches from the elaborate and beautiful drawings
of Mr. Scrope.

Montgy ; 3, Monjughat ; 4, Mont Dome in the distance.

Pl. IV. fig. 4, represents hills of secondary Jura limestone, capped by basalt (*coloured green*) ; the terminations of lava currents, that are extended over the whole area, covering a platform of primary rocks, and flowing on to the secondary, which, in this part of Ardêche, constitute an elevated limestone district.

This region affords a striking illustration of the erosion of the surface of a country by alluvial action. The thickness of the volcanic mass is between 300 and 400 feet; it is composed of two distinct beds of basalt, separated by a layer of scoriæ and volcanic fragments. Many portions, both of the upper and lower beds, are made up of well-defined, vertical, polygonal columns. The streams of lava to which these plateaux belong, have been traced by Mr. Scrope for more than thirty miles ; rising in a narrow ridge across the primitive heights, and then spreading over the secondary formations, and forming a parallelism with them. The limestone beneath the basalt is, in some places, covered with a vegetable soil, containing a common species of terrestrial shell (*Cyclostoma elegans*). The nearly horizontal disposition of the basalt, its columnar structure, and position on limestone, into which it has injected veins and dikes, render it, as Mr. Scrope observes, very analogous to the ancient volcanic rocks of Ireland, which I shall describe in a future lecture.

R

42. MONT DOR.—Before entering upon the description of the organic remains found in the rocks
and strata we have thus hastily surveyed, it will
be necessary to notice another system of extinct
volcanoes, situated in the same district, and connected with the Puy de Dome. While in the district I have just described, the primitive soil is only
partially obscured by the volcanic products, in
Mont Dor, the granitic foundation is covered over
an area of many miles in extent, and the erupted
masses attain a considerable elevation. Mont Dor
is a mountainous tract, the highest portion of which
is about 6,000 feet in altitude. It consists of seven
or eight rocky summits, grouped together within
a zone of a mile in diameter, the whole formed
of a succession of beds of volcanic origin. It is
deeply channelled by two principal valleys, and
furrowed by many minor water-channels, all having
their sources near the central eminence, and diverging towards every point of the horizon. The
beds of which this group is composed, consist of
scoriæ, pumice-stones, trachyte, and basalt; these
rocks all dip off from the central axis, and lie
parallel to the sloping flanks of the mountain, as is
the case in Etna, the Peak of Teneriffe, and all
other insulated volcanic mountains (see the section
of a volcano, Pl. IV. fig. 1). There is no crater, all
traces having been destroyed since the extinction
of its fires; but streams of lava may be traced in
elevated peaks, over a gorge which occupies the

very heart of the mountain. Some of these lava currents extend, as before stated, to a distance of many miles. A remarkable natural section, worn by a cascade, at a short distance from the baths of Mont Dor, exhibits the general structure of the beds (Pl. IV. fig. 5). They consist of—1. A bed of porphyritic trachyte; a volcanic rock, 160 feet in thickness. 2. Arenaceous tufa. 3. Columnar basalt. 4. Breccia, made up of volcanic fragments, cemented together by tufa. 5. Thick beds of basalt. 6. White ferruginous tufa, enveloping fragments of granite, basalt, &c. This bed is traversed by veins of the overlying basalt.

I will only remark, that the volcanic vents of central France are evidently of very different ages ; some being of immense antiquity, while others must be of comparatively recent origin, for they have exploded through the older beds of basalt.

43. FRESH-WATER LIMESTONE, AND ORGANIC REMAINS OF AUVERGNE.—The beds of the volcanic district of this province present alternations of limestone, abounding in fresh-water shells and other animal remains; with basalt, scoriæ, and other igneous productions, based on a foundation of granite. These deposites may be arranged in the following chronological order, beginning with the lowest or most ancient :—

1st. Clay, sand, and breccia, without organic remains.

2d. Limestone and calcareous marl, in strata nearly

R 2

horizontal; about 900 feet thick. These are
entirely of fresh-water origin, containing shells of
the genera (*Potamides, Helix, Planorbis,* and *Lymnea*) known to inhabit lakes and rivers. Some of
the beds contain bitumen ; others are entirely
made up of the cases of the caddis-worm (*Indusia
tubulata*), cemented together by calcareo-silicious
matter. This specimen, which was in the cabinet
of Faujas St. Fond, displays the characters of this
remarkable concrete : it consists of the tubes or
cases of the larva of a species of *Phryganea ;*
similar remains have been mentioned as occurring
in abundance in the alluvial silt of Lewes Levels
(page 88). The tubes are formed by the adhesion
of shells to the outer surface of the silken case
secreted by the insect; these cases are abandoned
by the animal when its metamorphosis is completed,
and groups of them may be seen in ditches or lakes.
In the fossil they have been cemented by calcareous
infiltration into a stone, so hard as to be employed
for building. The attached shells are so minute,
that more than a hundred are affixed to one tube,
and the space of a cubic inch often includes ten or
twelve tubes. If, says Mr. Scrope, we consider
that repeated strata, of five or six feet in thickness,
almost entirely composed of these tubes, were extended over the whole plain of the Limagne, occupying a surface of many hundred square miles, we
may have some idea of the countless myriads of
minute beings which lived and died within the

bosom of this once extensive lake. In this limestone, associated with land and fresh-water shells,
and remains of vegetables, are bones of the palæotherium, anoplotherium, lagomys, martin, dog,
rat, tortoise, crocodile, serpent, and birds, and in
which the lava current that has flowed over them
has produced but little alteration. This series of
strata comprises also beds of gypseous and laminated marls, with intercalations of silicious limestone, containing impressions of lake and. river
shells. In some localities, the fresh-water limestone has an intermixture of volcanic matter, presenting all the characters of a sediment slowly and
tranquilly deposited in a lake, into which ashes,
and fragments of rocks and scoriæ, were projected
by a neighbouring volcano ; while there are beds
which appear to have been formed by a violent intrusion of volcanic matter.

3d. Immense beds of volcanic production, consisting of basalt, scoriæ, &c. now existing in sheets
of lava, spread over the tabular masses of fresh-
water limestone, or as crests on the summits of the
lower hills. (Pl. IV. fig. 3.)

4th. Beds of sand and diluvial gravel, containing
bones of the mastodon, elephant, hippopotamus,
rhinoceros, tapir, horse, boar, felis, hyena, bear,
dog, castor, hare, &c. ; with these are associated
lignite, and other vegetable remains. Some of
the beds of limestone abound in the seed-vessels
of the Chara ; and the laminated clays contain

fishes, and leaves, stems, &c. of reeds and other plants.

There are several incrusting springs in Auvergne, largely impregnated with carbonic acid, which pour out immense quantities of calcareous tufa; these burst through the primitive rocks, which form the base of the whole territory, and cover the volcanic focus whence these mineral waters, in all probability, originate. Thermal springs are also very numerous throughout this volcanic district.

44. SUMMARY OF THE GEOLOGICAL PHENO-MENA OF AUVERGNE.—In the calcareous and sili-cious limestones of Auvergne, and their associated laminated marls, gypsum, lignite, and conglomerate, we have a general analogy with the older fresh-water tertiary formations of Paris and the Isle of Wight; the shells and plants being similar, and the quadrupeds of the same genera. And if we suppose the Paris basin to have been elevated during the active state of neighbouring volcanoes, and that successive streams of lava had poured over its sedimentary deposites, we should have a series of phenomena resembling those of Auvergne, with the exception that the presence of marine remains would denote that the basin had been filled with salt water. The facts submitted to our notice appear to establish the following sequence of physical events.

1st. The elevation, after the deposition of the secondary limestone, of the whole area of the primary rocks which form the foundation of central France.

2dly. A period of tranquillity, during which fresh-water lakes occupied the irregular hollows of the district; the neighbouring country being inhabited by Palæotheria, Anoplotheria, and other extinct mammalia, whose bones, together with the then existing vegetation, and the shells of the lacustrine mollusca, were imbedded in the tranquil depositions going on in the lacustrine basins.

3dly. Another elevation of the district; a new system of lakes was established, the country became clothed with forests, and inhabited by large deer, oxen, rhinoceroses, and hyenas; their remains being imbedded in the sediments of the waters.

4thly. The volcanoes became active; explosions took place through hundreds of vents; trachyte and basalt were ejected, piercing the fresh-water deposites in some places, and in others overspreading them with sheets of lavas. Vegetation still flourished, and remains of plants became entombed in the volcanic products.

5thly. Another period of tranquillity—the rivers, and other water-courses, dammed up, or deranged by the lava currents, formed new channels, and accumulated beds of gravel, sand, and clay. Gigantic deer and oxen still inhabited the district, having for contemporaries hyenas and other carnivora. Volcanic eruptions succeeded, and continued till a comparatively recent period.

Lastly. Floods and rivers of later date, which now constitute the drainage of the country, began to wear away channels through the beds of lava and limestone to the granite rock beneath, intersected the country with valleys and ravines, and spread over the ancient beds the modern alluvial soil.*

45. EROSION BY WATER-CURRENTS.—There is perhaps no district which exhibits in more striking characters the erosive power of running water, than Auvergne. In many places the basalt is columnar, like that of Staffa, and the Giants' Causeway; and one range, on the banks of the Ardêche, forms a majestic colonnade 150 feet in height, extending a mile and a half along the valley which has been channelled out by the stream that flows at its base. Mr. Scrope's description of this process is highly graphic. "The bed of the Ardêche is strewed with basaltic boulders, pebbles, and sand, originating from the destruction of the columnar ranges. In some of the volcanic cones the currents of basalt may be traced issuing from the water and following the inequalities of the valley, just as a stream of lava would flow down the same course at the present time. Yet these ancient currents have

* This account of the volcanic district of Auvergne, is an abstract of the interesting Essays of Messrs. Bakewell, Scrope, Lyell, Murchison, Dr. Daubeny, Dr. Hibbert, MM. Croiset, Jobert, Robert, and Bertrand-Roux. Mr. Scrope's work cannot be perused, even by the general reader, without deep interest.

since been corroded by rivers, which have worn
through a mass of 150 feet in height, and formed a
channel even in the granite rocks beneath, since
the lava first flowed into the valley. In another
spot, a bed of basalt 160 feet high, has been cut
through by a mountain stream, and very beautiful
columnar masses are displayed. The vast excavations effected by the erosive power of currents along
the valleys which feed the Ardèche, since their
invasion by lava currents, prove that even the
most recent of these volcanic eruptions belong to
an era incalculably remote."

46. EXTINCT VOLCANOES OF THE RHINE.—
I have dwelt so long on the Phlegræan fields of
Auvergne, that but a brief space can be afforded
to another group of tertiary volcanoes. Every one
who has ascended the Rhine, will remember where

> " The castled crag of Drachenfels
> Frowns o'er the wide and winding Rhine,"

forming one of the Siebengebirge, or Seven Mountains, whose majestic and graceful forms suddenly
burst on the sight, rising from the level plains on
the right bank of the river, to an altitude of nearly
1,500 feet. These picturesque objects form part of
a group of extinct volcanoes; while, on the opposite
side of the river, the Eifel, with its crater covered
with scoriæ and cinders, and lava currents, still distinctly visible, attest the wide area over which these
ancient fires once extended. Unlike the district

we have just noticed, the foundation-rock of the
country is an ancient sedimentary deposit, called
greywacke, consisting of coarse red sandstone and
slate of a peculiar character, which we shall describe
hereafter, thrown into a highly inclined position.
Through these beds the volcanic eruptions, consisting of trachyte, basalt, and other modifications
of trap rocks and scoriæ, have forced their way.
The basalt is black, very compact, and breaks
into sharp fragments; it is frequently columnar,
and the separate hexagonal pillars are made use
of for posts, and paving, in the adjacent towns.
Such, says Mr. Horner (whose admirable memoir*
has furnished the materials for this imperfect
sketch), is the profusion of basaltic pillars, that
the walls of the town of Linz are wholly built
of these materials, placed on their sides, with
the ends projecting outwards.   The streets are
paved with the smaller columns set on end, thus
forming a miniature representation of the Giants'
Causeway ; and the same volcanic product forms a
large proportion of the walls of Bonn and Cologne.
The greywacke is covered by a series of tertiary
deposites, consisting of beds of sand, sandstone, clay,
and lignite, constituting what is termed a *brown
coal formation.*   On these strata an extensive layer
of gravel is superposed, over which is spread a
loosely coherent, sandy loam, provincially termed

* On the Geology of the Environs of Bonn, by Leonar
Horner, Esq. F.R.S.   Geological Transactions, vol. iv. 1836.

Loess, containing recent species of terrestrial and fresh-water shells, and forming the subsoil of the vast plains in which Bonn and Coblentz are situated, extending as far as the falls of Schaffhausen.

47. BROWN COAL FORMATION.—As the usual condition in which bituminized vegetable matter occurs in the tertiary formations, is well exemplified in the brown coal, or lignite, of the Rhine, it will be instructive to examine the characters of this deposit somewhat in detail; for we shall thereby obtain data which will prepare us for the investigation of the ancient carboniferous deposites. This formation, which is spread over a great extent of country on both sides of the Rhine, consists of layers of clay, sand, sandstone, conglomerates, clay and ironstone, and lignite, or bituminized wood of various qualities, disposed in distinct beds, and intermixed with argillaceous matter. The breadth of the ridge of low hills formed by this assemblage of deposites on the left bank of the Rhine, is from three to five miles, its elevation varying from 50 to 200 feet.

The lignite occurs in the following states:—
1. A black earthy and pulverulent substance. 2. Concretionary masses, with leaves and fragments of wood. 3. Wood in various degrees of bituminization, and of shades of colour, from a light-brown to jet-black. 4. Very finely laminated masses of bituminous matter and clay, of a dark chocolate colour, and separating into elastic flakes, as thin as

paper, whence its name *Papierkohle.* These specimens, collected by Sir P. M. de Grey Egerton, Bart. exhibit the peculiar character of the substance, which is so highly bituminous as to burn with a bright flame. The wood is generally in inconsiderable fragments; but stems of large trees, somewhat compressed, occasionally occur; in some instances the trees are imbedded in an upright position, with the roots attached and the stems passing through several beds of lignite. In many examples the wood is so little changed, that, like the timber of our peat-bogs, it is employed in building; in others it is highly pyritous—in other words, is impregnated with sulphuret of iron, like the fossil vegetables of Sheppey. Mr. Horner states, as the result of his investigations, that there were extensive fresh-water lakes, in the sediments of which trees and plants, drifted by land-floods, were engulfed; and that volcanic eruptions were simultaneously going on, in the same manner as in the modern submarine volcanoes. There is a great fault, or dislocation, of the brown coal formation, which may be attributed to a powerful and sudden volcanic explosion, that probably occasioned the elevation of the Siebengebirge, and raised up that portion of the coal-beds which repose on the flanks of these peaks. The gravel covering the lignite, must have been strewed over the plain previous to this elevation, for it is found on both sides of the river at a great height, and not in the intermediate plain.

These inductions are so evident as to require no comment.

The ancient alluvium called the *Loess*, very much resembles a bed of loam which occurs in some parts of Lewes Levels, and incloses fresh-water and land shells of many existing species. The Loess rarely contains bones of quadrupeds; but Mr. Lyell mentions that remains of the horse and mammoth have been discovered.* From the extensive distribution of this deposit, and its occurrence at various elevations, in some instances on the flanks of mountains 1,200 feet above the level of the sea, at others spread out over the gravel of the vast plain of the Rhine, Mr. Lyell infers, that although the *Loess* has been deposited since the existing system of the hills and valleys of the country, yet that changes must have taken place in the physical geography of the district, subsequent to its original formation; and that there is reason to conclude, that since the deposition of this fluviatile loam all the land between Switzerland and Holland has suffered a subsidence, and a subsequent elevation, to the amount of many hundred feet.†

48. OTHER TERTIARY DEPOSITES OF EUROPE, NORTH AMERICA, &c. — It has already been mentioned that strata, referrible to the period compre-

---

* Principles of Geology, vol. iv. p. 33.

† Ibid; read with particular attention, pp. 36 & 37.

hended between the newest secondary formations and the human epoch, occur throughout Europe, presenting in some instances well-defined groups, with marked boundaries; in others, vast areas, over which these deposites are irregularly spread. The geographical relations of the tertiary strata to the existing lands and seas, is an inquiry of deep interest, but one into which my limits forbid me to enter. I will only remark, that Europe must have possessed many of its most striking physical characters at the commencement of the Eocene period; and that its present configuration has been produced by the conjoint effect of successive mutations in the relative level of land and water, during the formation of the marine and fresh-water deposites, reviewed in this discourse. In India formations of like character have been observed in the Burmese empire, in the Sub-Himalaya mountains, and in the Caribari hills; and among the remains of various quadrupeds a new species of Anthracotherium, one of the genera discovered in the tertiary strata of France, has been found.

In North America the researches of Dr. Morton, Professor Vanuxem, and other observers, have shown that in the territories of the United States tertiary deposites extend over a great part of Maryland, along the coast of New York and New England, and occur in New Jersey, Delaware, Long Island, &c. The tertiary beds of Maryland consist of limestone, clay, sand, and gravel, and

abound in the usual types of European tertiary
marine shells. I have placed before you an extensive collection from the United States, for which I
am indebted to the kindness of Dr. Morton, Mr.
Conrad, Professor Silliman, Dr. Harlan, and other
American *savans;* and you will observe how striking is the general analogy between these shells and
those of the Paris and London basins. The *Turri-
tella, Venericardia, Fusus, Ancilla,* &c. are identical
with the European species; but some of the types
are altogether new.

49. ALTERED TERTIARY STRATA OF THE ANDES.
—But striking as are the proofs already adduced of
elevations, and other effects of volcanic agency,
during the tertiary period, these sink into comparative insignificance when contrasted with the
enormous changes which have taken place in the
great mountain chains of South America during
the same geological cycle. From the researches of
an eminent naturalist and highly intelligent observer, Mr. Charles Darwin, we learn that an
extensive tertiary system, analogous to those of
Europe, skirts both flanks of the primary rocks
which form the southern chain of the Andes, the
latter having suffered a certain degree of elevation
before the deposition of the former. The tertiary
strata are of great thickness and extent, and
separable into two groups; the lowermost beds,
like those of Auvergne, repeatedly alternate with
lavas, and thus denote the commencement of the

eruptions of the ancient craters. Over these deposites are accumulations of porphyritic pebbles,
covered, at elevations of many hundred feet, by
beds of shells of recent species; and the sides
of the mountains appear like a succession of
lifted sea-beaches, which have been slowly and
tranquilly lifted up during a long period of time.
The altered character of the tertiary deposites
within the influence of the igneous products,—the
transmutation of accumulations of loose pebbles
into solid, compact rocks,—and the occurrence of
metalliferous veins in strata of comparatively modern
formation—are facts so powerfully exemplifying the
geological principles enunciated in the former
lectures, that although this discourse has extended
to a great length, I cannot omit Mr. Darwin's
spirited and graphic description of these phenomena, as originally communicated in a letter to
Professor Henslow, of Cambridge, dated Valparaiso,
March 1835.

" You will have heard of the dreadful earthquake of the 20th
February. I wish some of the geologists, who think the earthquakes of these times are trifling, could see how the solid rocks
are shivered. In the town there is not one house habitable; the
ruins remind me of the drawings of the desolated eastern cities.
We were at Valdivia at the time, and felt the shock very se-
verely. The sensation was like that of skating over very thin
ice, that is, distinct undulations were perceptible. The whole
scene of Conception and Talesana is one of the most interesting
spectacles we have beheld since we left England. I was much
pleased at Chiloe by finding a *thick bed of recent oyster-shells, &c.
capping the tertiary plain,* out of which grew large forest trees.

I can now prove that both sides of the Andes have risen in this recent period to a considerable height. Here the shells were 350 feet above the sea. On the bare sides of the Cordilleras complicated dykes and wedges of variously coloured rocks are seen traversing, in every possible form and shape, the same formation, and thus proving by their intersections a succession of violences. The stratification in all the mountains is beautifully distinct, and owing to a variety of colouring can be seen at great distances. Porphyritic conglomerates, resting on granite, form the principal masses. I cannot imagine any part of the world presenting a more extraordinary scene of the breaking up of the crust of the globe than these central peaks of the Andes. The strata in the highest pinnacles are almost universally inclined at an angle from 70° to 80°. I cannot tell you how much I enjoyed some of the views; it is alone worth coming from England to feel at once such intense delight. At an elevation of from ten to twelve thousand feet there is a transparency in the air, and a confusion of distances, and a stillness, which give the sensation of being in another world. The most important and most developed formation in Chili is the porphyritic concrete. From a great number of sections I find it to be a true coarse conglomerate or breccia, which passes by every step in slow gradation to a fine clay-stone porphyry; *the pebbles and cement becoming porphyritic, till at last all is blended in one compact rock.* The porphyries are excessively abundant in this chain, and at least four-fifths of them, I am sure, *have been thus produced from sedimentary beds in situ.* The Uspellata range is geologically, although only six or seven thousand feet high, a continuation of the grand eastern chain. It has its nucleus of granite, consisting of beds of various crystalline rocks, (which I have no doubt are subaqueous lavas,) alternating with sandstone, conglomerates, and white aluminous beds, like decomposed felspar, with many other curious varieties of sedimentary deposits. In an escarpment of compact greenish sandstone, *I found a small wood of petrified trees in a vertical position,* or rather the strata were inclined about 20° or 30° to one point of the trees, 70° to the other; that is, before the tilt, they were truly vertical. The

S

sandstone consists of many horizontal layers. Eleven of the trees are perfectly silicified, and resemble the dicotyledonous wood which I found at Chiloe and Conception; the others, from thirty to forty in number, I only know to be trees from the analogy of form and position; they consist of snow-white columns of coarsely crystallized carbonate of lime. The largest trunk is seven feet in circumference. They are all close together, within one hundred yards, and about the same level; no where else could I find any. It cannot be doubted that the layers of fine sandstone have quietly been deposited between a clump of trees, which were fixed by their roots. *The sandstone rests on lava; is covered by a great bed, apparently about one thousand feet thick, of black augite lava; and over this there are at least five grand alternations of such rocks, and aqueous sedimentary deposits, amounting in thickness to several thousand feet.* According to my view of these phenomena, the granite, which forms peaks of a height probably of 14,000 feet, has been fluid in the tertiary epoch; strata of that period have been altered by its heat, and are traversed by dykes from the mass, and are now inclined at high angles, and form regular or complicated anticlinal lines. *To complete the climax, these same sedimentary strata and lavas are traversed by very numerous true metallic veins of iron, copper, arsenic, silver, and gold, and these can be traced to the underlying granite. A gold mine has been worked close to the clump of silicified trees!*"

50. TERTIARY SALIFEROUS DEPOSIT.—Not only coal, but even extensive deposites of rock salt occur in the tertiary system. The celebrated salt mines of Gallicia, of which my friend M. Boué* has given an interesting description, are referrible to this epoch. The deposit is nearly 3000 yards long, 1066 broad, and 280 yards deep. The upper part

---

* *Journal de Géologie,* quoted by M. De la Beche.

of the mine consists of green salt, with nodules of gypsum in marl. The salt contains in some places lignite, bituminous wood, and shells. In the lower division are beds of arenaceous marls, with lignite, impressions of plants, and veins of salt; coarse sandstone, with vegetable remains; aluminous and gypseous shale, and indurated calcareous marl, with sulphur, salt, and gypsum.

51. RETROSPECT.—So numerous and varied have been the phenomena presented to our notice in this general survey of the tertiary strata, that a comprehensive retrospect is necessary, in order that we may arrive at the important inferences which they offer.

In the pliocene, or newer tertiary, which also embraces the mammalian epoch of the last lecture, the fossil remains in the alluvial deposites afford incontestible proof that the mammoth, mastodon, hippopotamus, dinotherium, and other colossal animals of extinct genera or species, together with birds, reptiles, and enormous carnivora, inhabited such districts of our continents as were then dry land; while the older tertiary, or eocene, inclose the bones of land animals, principally of a lacustrine character, which approximate to certain races that now exist in the torrid zone, but belong to extinct genera, which preceded the mammoth and the mastodon. The seas and lakes of that remote epoch occupied areas which are now above the waters; and rocks and mountains, hills and valleys, streams and rivers, diversified the surface of countries which are now destroyed or entirely changed; and whose past existence is revealed by the spoils which the streams and rivers have accumulated in the ancient lakes and seas. The ocean abounded in mollusca, crustacea, and fishes, a large proportion of which is referrible to extinct species. Crocodiles, turtles, birds, and insects, were contemporary with the palæotherium, and anthracotherium; and animal organization, however varied in certain types, presented the same general outline as in modern times; the extinction of species and genera being then, as now, in constant activity. The vegetable world also contained the same great divisions; there were forests of oak, elm, and beech; of firs, pines, and other coniferous trees; palms, tree-ferns, and the principal groups of modern floras; while the water, both salt and fresh, teemed with the few and simple forms of vegetable structure peculiar to that element. The state of the inorganic world is not less manifest: the abrasion of the land by streams and rivers,—the destruction of the sea-shores by the waves, and the formation of beach and shingle,—the desolation inflicted by volcanic eruptions,—all these operations were then, as now, in constant action. The bed of an ancient sea, containing myriads of the remains of fishes, crustacea, and shells, now forms the site of the capital of Great Britain; and the accumulation of tropical fruits and plants, drifted by ancient currents from

other climes, constitute islands in the estuary of
the Thames; while the sediments of lakes and gulfs,
teeming with the skeletons of beings which are
blotted out from the face of the earth, make up the
soil of the metropolis of France.

Although the changes in the relative level of the
land and sea during this epoch were numerous and
extensive, yet one region still preserves traces of
its original physical geography, and although the
earthquake has rent its mountains to their very
centre—though hundreds of volcanoes have again
and again spread desolation over the land—and
inundations and mountain torrents have excavated
valleys, and chequered the plains with ravines and
water-courses—yet the grand primeval features of
that country remain; and we can trace the boundaries of its ancient lakes, and the succession of
changes it has undergone from the first outbreak
of its volcanoes, to the commencement of the present
state of repose. The lowermost lacustrine deposites
in Auvergne, which are spread over the foundation
rock of granite unmixed with igneous productions,
mark the period antecedent to the volcanic era;
while the intrusions of lava and scoriæ in the superincumbent strata, denote the first eruptions of
Mont Dome. The succeeding period of tranquillity
is recorded in characters alike intelligible. The
slow deposition of calcareous mud—the incrustation
of successive generations of aquatic insects, crustacea, and mollusca, and we may even add of infusoria

—the imbedding of the bones of mammalia, birds, and reptiles—the accumulation of lignite and other vegetable matter—are data from which we may restore the ancient country of Central France. It was a region encircled by a chain of granite mountains, and watered by numerous streams and rivulets, and possessing lakes of vast extent. Its soil was covered with a luxuriant vegetation, and peopled by palæotheria, anoplotheria, and other terrestrial mammalia; the crocodile and turtle found shelter in its marshes and rivers; aquatic birds frequented its fens, and sported over the surface of its lakes; while myriads of insects swarmed in the air, and passed through their wonderful metamorphoses in the waters. In a neighbouring region,* herds of ruminants and other herbivora, of species and genera now no more, with birds and reptiles, were the undisturbed occupants of a country abounding in palms and tree-ferns, and having its rivers and lakes, with gulfs which teemed with the inhabitants of the sea; and to this district the fiery torrents of the volcano did not extend. But to return to Auvergne—a change came over the scene—violent eruptions burst forth from craters long silent—the whole country was laid desolate—its living population swept away—all was one vast waste, and sterility succeeded to the former luxuriance of life and beauty. Ages rolled by—the mists of the mountains

* The Paris basin is about 220 miles from Auvergne.

and the rains, produced new springs, torrents, and rivers—a fertile soil gradually accumulated over the cooled lava currents and the beds of scoriæ, to which the sediments of the ancient lakes, borne down by the streams, largely contributed. Another vegetation sprang up—the mammoth, mastodon, and enormous deer and oxen now quietly browsed in the verdant plains—other changes succeeded— these colossal forms of life in their turn passed away, and at length the earlier races of mankind took possession of a country, which had once more become a region of fertility and repose.

To those who have favoured me with their attention through these discourses, it cannot be necessary to insist that the changes in organic and inanimate nature, which I have thus rapidly portrayed, are supported by proofs so incontrovertible, and traced in language so intelligible, as to constitute a body of evidence with which no human testimony can compete. It is true that the time required for this succession of events must have extended over an immense period; but, as I have before remarked, time and change are great only in relation to the beings which note them, and every step we take in geology, shows the folly and presumption of attempting to measure the operations of nature by our own brief span. "There are no minds," says Mr. Scrope, " that would for one moment doubt that the God of Nature has existed *from all eternity*; but there are many who would reject as preposterous, the

idea of tracing back the history of *His works* a
million of years. Yet what is a million, or a million
of millions of years, when compared to eternity ?"*

Germany presents us with an interesting series
of analogous changes, effected in a later era. The
outburst of the now extinct volcanoes of the Rhine,
the accumulation of fluviatile silt over the plains,
and the subsequent elevation of the whole country,
show that these physical mutations were not confined to a single region or period.

In the Andes, the enormous disruptions and
elevations of the most ancient as well as modern
deposites, teach us, that through a long lapse of
ages, the volcanic fires of South America have acted
with intense energy ; and yet more, that the melting
and transmutation of loose materials into compact
rocks, the conversion of incoherent strata into solid
stone, and even the sublimation of gold and other
metals into fissures and veins, are phenomena
which have taken place since our seas were peopled
by existing species of mollusca. The importance of
these extraordinary and interesting facts, will be
rendered more obvious in a subsequent lecture.

In conclusion ; it will be useful to inquire, even
though some repetition may be incurred, what are
the legitimate inferences as to the condition of the
earth and its inhabitants during the tertiary epoch,
from the facts that have been placed before us ?

* Geology of Central France.

Was there, as some have supposed, an essential
difference in the constitution of the earth? —
was its surface more covered with lakes and
marshes than now? — and did animal life more
abound in those types, which are suitable to a
lacustrine condition? — or have these conclusions
been drawn from a partial view of the phenomena,
and do the facts only warrant the inference that
certain regions which are now dry land were in
ancient times occupied by vast lakes, and that there
may have existed contemporaneously as great an
extent of dry land as at present, in areas now buried
beneath the ocean? In the fossilized remains of
the tertiary population of the land and waters, we
find all the grand types of the existing animal
creation—terrestrial, lacustrine, and marine mammalia—herbivora, carnivora, birds of every order,
and of numerous species and genera—reptiles, fishes,
crustacea, insects, zoophytes, and even those living
atoms, the infusoria*—in short, all the leading divisions, and even sub-divisions of animal existence.
In the vegetable world, as I have already remarked,
the same general analogy is maintained. And as
all these varied forms of being required physical
conditions suitable to their respective organizations,
we have at once conclusive evidence that the general constitution of the earth in the tertiary epoch,

* In tertiary strata the remains of beings of which millions
would occupy but a cubic inch have been discovered. They will
be described hereafter.

could not essentially have differed from the present.
Dry land and water, continents and islands, existed
then as now—their geographical distribution may
have varied—the temperature in certain latitudes
may have been much higher—countries may have
existed in areas now covered by the water, and
marshes and fens have prevailed in regions now
arid and waste ; but the same agents of destruction
and renovation were then, as now, in constant
activity. It is true that immense numbers of large
mammalia lie buried in regions where it is utterly
impossible such creatures could now find subsistence, and in latitudes whose climates are unsuitable
to such forms of organization. But some of these
apparent anomalies may be explained by the fact,
that the alluvial beds in which these remains occur,
cannot have been the sites of the dry land on which
these lost beings existed; they are the sediments
of ancient lakes—the deltas of former rivers—the
estuaries of seas—they are formed of the detritus of
the land transported from a distance. If the Gulf-
stream annually strews the shores of the Hebrides
with the fruits of torrid climes, the currents of the
ancient seas must have produced analogous results ;
and in our attempt to interpret past changes, it
must not be forgotten that they have most probably
been produced by causes which are still in action.
I do not question the assumption that the countries containing these fossil remains, may have
enjoyed a milder climate in the tertiary epoch than

at present; or that in still earlier periods there may not have prevailed all over the world a higher temperature than now. But it appears to me that the variation of climate which a change in the distribution of the land and water would occasion, as suggested by Mr. Lyell,—or a difference in the radiation of heat from internal sources, as explained by Sir J. Herschel, and Mr. Babbage,—may account for these phenomena.

The occurrence of groups of animals of the same families, in certain districts, is in strict conformity with the distribution of living species, in regions not under the control of man; and thus when ancient France presented a system of lakes, animals fitted for such physical conditions found there the means of subsistence—when the vast plains and forests of America were adapted for colossal mammalia, there the mastodons and the mammoths obtained food and shelter—and when the former continent of Europe swarmed with herbivora, the carnivorous tribes, as the lion and the tiger, the bear and the hyena, obtained the support which their habits and economy required.

One striking feature in the events that have passed in review before us, is the immense scale on which the extinction of species and genera has been effected : but it must be remembered that our observations have extended over a period of vast duration, and that we therefore have seen the aggregate effects of a law, which even before our eyes is producing great and important modifications in the system of animated nature.

Thus the tertiary epoch displays to us a state of the earth replete with life and happiness : the physical constitution of its surface being then, as now, admirably adapted to the habits and economy of the beings it was designed to support. In the most ancient periods, forms of life prevailed which gradually became extinct, and were succeeded by others which in their turn also passed away ; and if we trace the varying types of being from the earliest ages, we perceive a gradual approach to the present condition of organic existence ; the grand line of separation between the present and the past being the creation of the human race. From that period, in proportion as man has extended his dominion over the earth, many races of animals have been either exterminated, or modified by his caprices or necessities; and it cannot be doubted that in the lapse of a few thousand years, a total change will have been effected by human agency alone, in the relative numerical proportions of existing genera and species.

LECTURE IV.

1. Introductory Remarks. — The knowledge we have acquired from our investigation of the phenomena described in the previous lectures, will materially facilitate our geological progress, by enabling us to comprehend the former effects of those agencies, by which, through all time, the surface of the earth has been renovated and maintained.

The elevation of the beds of seas and rivers, and their conversion into fertile countries—the submergence of islands and continents beneath the waters of the ocean—the rapid formation of conglomerates from shells and corals on the sea shore—the accumulation of beach and gravel, and the inhumation of animals and vegetables—the slow deposition of sediment by lakes, the imbedding of countless generations of insects, and the formation of limestone from their almost invisible skeletons—the construction of solid stone out of fragments of bones, and rocks, shivered by earthquakes—the engulfing, in estuaries and inland seas, of land animals, birds, and reptiles—the consolidation of both organic and inorganic substances into rock, by the infiltration of flint and lime by thermal waters—the transmutation of immerged forests into coal and lignite—the destructive and conservative effects of volcanic eruptions—the conversion of sand, gravel, and clay into homogeneous masses by heat, and even the production of metalliferous veins of gold and silver—all these phenomena have passed in review before us, although our inquiries have extended but through periods which, however vast and remote in relation to the records of our race, are brief and modern in the physical history of the earth.

The geological events previously described, although forming a connected series, may be divided into periods, each marked by certain zoological characters. 1st. The modern, or human epoch;

2d. the elephantine, characterised by the preponderance of large pachydermata ; 3d. the palæotherian, in which animals allied to the tapir prevailed, and Europe presented a system of gulfs and lakes.

2. SECONDARY FORMATIONS.—I hasten to the consideration of another geological epoch,—that which comprehends the Secondary Formations. Hitherto our attention has been principally directed to deposites confined within comparatively limited areas ; the basins of lakes, gulfs, estuaries, and inland seas ; the drifted accumulations of torrents, rivers, and inundations. We have now arrived on the shores of that ocean, of whose spoils the existing islands and continents are principally composed ; the fathomless depths of the ancient seas are spread before us, and all the myriads of beings which sported in their waters, and lived and died in those profound abysses, remain, like the mummies of ancient Egypt, the silent yet eloquent teachers of their own eventful history.

A reference to the Tabular Arrangement of the Strata (Pl. III. page 178), will show that the secondary formations constitute eight principal groups. I propose in this discourse, to explain the geological characters of the first two in the series, namely, the CHALK and the WEALDEN. The former is composed of strata that have been accumulated in the depths of a sea of great extent; the latter, of the sediments of an ancient delta; the one affording a striking illustration of the nature of *oceanic*, and the other of *fluviatile* deposites.

In the diagram, Plate III., the Wealden (3*) is represented as an intercalation between the Chalk and the Oolite (3, 4), because it is of limited extent, and where absent, as in the midland counties of England and on the continent, the chalk˅lies upon the oolite, as will be shown in the next lecture. As both the chalk and the wealden are fully developed in the south-east of England, the phenomena about to be described may be examined with but little inconvenience ; and an extensive collection of their peculiar fossils may be seen in my museum.*

3. THE CHALK FORMATION.—The pure white limestone, called *Chalk*, is known to every one ; but in the nomenclature of geology, the term is applied to a group of deposites very dissimilar in their lithological compositions, but agreeing in the character of the organic remains which they contain, and evidently referrible to the same epoch of formation. With this explanation, it will be convenient to employ the term in its extended sense.   The Chalk formation is composed of four principal divisions.   1. The uppermost consists of *Chalk*, with and without flints.   This limestone is generally white, but in some districts red, and in others yellow ; flint nodules and veins abound in the upper, but seldom occur in the lower division.   2. *Marl*,

* See Descriptive Catalogue of the Mantellian Museum, 8vo.

an argillaceous limestone, which universally prevails beneath the white chalk; it sometimes contains a large intermixture of green sand, and then forms what is called *Firestone*, or *Glauconite*. 3. *Galt*, a stiff, blue or black clay, abounding in shells, which frequently possess a pearly lustre. 4. *Shanklin*, or *Green Sand*, a triple alternation of sands and sand-stone with clays; beds of chert and fuller's earth are found in some localities.

On the Continent, and in other parts of the world, indications of the series of deposites here enumerated, occur; and, taken as a whole, the chalk formation may be described as extending over a great part of the British Islands, Northern France, Germany, Denmark, Sweden, European and Asiatic Russia, and of the United States of North America.* Over this vast extent, the organic remains of the chalk maintain certain general characters, sufficiently obvious to determine the nature of the formation.

Whether imbedded in pure white limestone, coarse sandstone, blue clay, loose sand, or compact rock, the fossils consist of the same species of shells, corals, sponges, echinites, belemnites, ammonites, and other marine exuviæ; fishes, reptiles, wood, and plants. The strata are well displayed along the Sussex and Kentish coasts, and those natural sections exhibit the manner in which the beds have

* Dr. Morton's Synopsis of the Cretaceous Groups of the United States, 1 vol. 8vo. with plates.

T

been disrupted, and thrown into an inclined position.
(See Pl. V. fig. 1.) Near Devizes, in Wiltshire,
the strata lie nearly horizontal, and in the following
order :—1. White Chalk. 2. Glauconite. 3. Galt.
4. Shanklin Sands. (See Pl. V. fig. 2.)

4. CHALK AND FLINT.*—The *white chalk* is
composed of lime and carbonic acid, and may have
been precipitated from water holding lime in solution, from which an excess of carbonic acid was
expelled. But some masses of chalk are composed
of minute corals and shells, and whole layers, in
many quarries, are formed of the *ossicula* of star-fish
and other radiaria. The nodules and veins of flint
which occur in the chalk, show that water holding
silex in solution must have been very abundant at
the cretaceous period. The power possessed by
thermal waters of dissolving silicious earth, depositing flint, and occasioning the silicification of
vegetable substances, is strikingly exemplified in
the Geysers of Iceland, as I have already explained
(p. 74). The perfect fluidity of the silex before
consolidation, is shown by the sharp impressions
which the flints bear of shells and other marine
bodies ; and upon breaking the nodules, sponges,
alcyonia, and other organic remains, are found

* See The Fossils of the South Downs, or, Illustrations of
the Geology of Sussex, 1 vol. 4to. with 42 plates. Geology
of the South-East of England, 1 vol. 8vo. Dr. Fitton's Memoirs on the Shanklin, or Green Sands, in which many of the
fossils are beautifully delineated.

enveloped, the silicious matter having so penetrated the delicate structure of the original, that polished sections display the minute organization of the inclosed zoophytes.

5. FLINT NODULES.—Flints, or silicious nodules, occur in the chalk in horizontal rows, which present some degree of regularity, and are placed at unequal distances from each other. This arrangement has probably arisen from the chalk and flint having been held in suspension or solution in the same fluid, and erupted into the basin of the ocean: when consolidation took place, the silicious molecules separated from the cretaceous, on the well-known principles of chemical affinity; the sponges and other zoophytes acting as nuclei or centres, around which the silicious matter coagulated. This process receives illustration from the fact, that when different substances in a state of extreme division are mixed together, they have a tendency to separate, and re-arrange themselves in masses more nearly homogeneous; thus a separation of pounded flint from aluminous earth, in the materials prepared in the potteries, will take place, and silicious concretions be formed, if the mixture be not constantly agitated. The marked stratification of the chalk shows that it was poured out periodically; and it is not unusual to find veins of flint running through and filling up crevices in the strata beneath; an appearance that can only be attributed to the lower beds having been consolidated, and subsequently

fissured, before the superincumbent stratum was precipitated.

6. SULPHURET OF IRON.—Iron pyrites is the only metalliferous ore that occurs abundantly in the chalk of England. The large nodular masses that are found on the Downs and in the ploughed fields, are commonly termed *thunderbolts*. This mineral is sometimes found in octahedral crystals of great elegance and regularity, and frequently occupies the cavities of shells, and echini; terebratulæ, and pectens, also occasionally occur in masses of this substance. The bones and scales of the fishes are invariably coloured with a ferruginous stain, arising from a curious chemical process; sulphuretted hydrogen was evolved during the course of putrefaction, but the sulphur entering into combination with the iron contained in the water, sulphuret of iron has formed, and thus the fossil fishes have derived the rich colour which so beautifully contrasts with the white chalk by which they are surrounded.

7. ST. PETER'S MOUNTAIN, MAESTRICHT.— I have described the usual lithological characters of the chalk, and if our observations were restricted to the deposites as they occur in England, the difference between the uppermost *secondary* formation and the superimposed *tertiary* would be most striking, both as regards the nature of the rocks and their organic remains. But, as I shall hereafter explain, the chalk of England appears to have been

formed in the most profound depths of the sea, and
we have rarely any intermixture of terrestrial or
littoral productions; even pebbles are of unfre-
quent occurrence. At Castle Hill, near Newhaven
(p. 208), and at Alum Bay, in the Isle of Wight
(p. 206), the *cerithia* (Tab. XXII. fig. 4), *turritellæ*,
and other tertiary shells, abound in the sand and
clay spread over the surface of the chalk, in which
no similar shells can be detected. On the Continent, however, there exist deposites which form a
link between the tertiary and secondary; and in
the valley of the Meuse a fine series of strata,
the uppermost abounding in those genera of shells
so plentiful in the tertiary, and passing impercep-
tibly into limestone with cretaceous fossils, and
flint nodules, and finally into chalk. The quarries
of St. Peter's Mountain have long been celebrated
for their remarkable fossils; but the true geological
characters of the strata were first determined by
the able investigations of Dr. Fitton. In North
America the researches of Dr. Morton have proved
the existence of analogous deposites.

St. Peter's Mountain, in which the quarries are
situated, is a cape or headland between the Meuse
and the Jaar, and forms the extremity of a range of
hills which bounds the western side of the valley
of the Meuse. The beds of limestone present a
total thickness of 500 feet. Excavations have for
centuries been carried on, and from the immense
quantities of stone removed, extensive caverns and

galleries now traverse the heart of the mountain.* Shells, corals, crustacea, teeth of fishes, and other marine remains, are in profusion; with wood perforated by lithodomi, and the bones of a large and very remarkable reptile.

8. THE MOSÆSAURUS; OR FOSSIL REPTILE OF MAESTRICHT.—The large bones and teeth of an unknown animal which were occasionally found in the limestone, had long since directed the attention of naturalists to the quarries of St. Peter's Mountain. In 1770 M. Hoffmann, who was forming a collection of organic remains, had the good fortune to discover a specimen, which has conferred additional interest on this locality: some workmen, on blasting the rock in one of the caverns of the interior of the mountain, perceived, to their astonishment, the jaws of an enormous animal attached to the roof of the chasm. The discovery was immediately made known to M. Hoffmann, who repaired to the spot, and for weeks presided over the arduous task of separating from the rock the mass of stone containing the remains. His labours were at length repaid by the successful extrication of the specimen, which he conveyed in triumph to his house. Unfortunately, the canon of the cathedral, which stands on the mountain, claimed the fossil in right of being lord of the manor, and succeeded by a troublesome and expensive law-suit

* See Hist. Nat. de la Montagne de St. Pierre, by Faujas St. Fond, 1 vol. 4to. with splendid engravings.

in obtaining this precious relic.   It remained in his
possession for years, and Hoffmann died without
regaining his treasure, or receiving any compen-
sation.   The French Revolution broke out, and the
armies of the Republic advanced to the gates of
Maestricht;   the town was bombarded, but by
desire of the committee of *savans,* who accompanied
the French troops, the artillery were not allowed
to play on that part of the city in which the celebrated fossil was known to be contained.   In the
meanwhile the canon, shrewdly suspecting why
such peculiar favour was shown to his residence,
concealed the treasure in a secret vault; but when
the city was taken the French authorities compelled
him to give up his ill-gotten prize, which was immediately transmitted to the *Jardin des Plantes,* at
Paris, where it still forms one of the most striking
objects in that magnificent collection.   It is but
just to add, that the relatives of Hoffmann were
rewarded by the French Commissioners.   The model
of this specimen in my museum was presented to
me by Baron Cuvier; it consists of the jaws, teeth,
palate-bone, vertebræ, and *os quadratum,* a bone possessed by some reptiles, and in which the auditory
cells are contained.   There are portions of jaws
with teeth of the mosæsaurus in the British Museum.
The original was a reptile, holding an intermediate
place between the *Monitor* and *Iguana,* about
twenty-five feet long, and furnished with a tail of
such construction as must have rendered it a power-

ful oar, enabling the animal to stem the waves of
the ocean, of which Cuvier supposes it to have been
an inhabitant. The vertebræ before you belong to
this creature, and were discovered in a chalk quarry
near Lewes. This remarkable specimen was found
a few days since, in the chalk of Kemp-town; it
is a vertebra of the tail, partially invested with
flint, which has consolidated around it without
obscuring its essential characters. These teeth,
from North America, collected by Dr. Morton,
appear to be identical with those from Maestricht,
and afford proof of the original extension of the
ocean of the chalk over the area now occupied by
the Atlantic.

9. ORGANIC REMAINS OF THE CHALK.—The
fossils of the chalk are very numerous, and comprise all the usual forms of marine existence, with
the exception of cetacea. Certain genera and species
appear restricted to certain subdivisions of the
formation. Thus in the white chalk, there are many
species of shells that do not occur in the other
divisions of the strata. The marl and galt are also
characterised by peculiar forms, and the Shanklin
sands abound in shells and zoophytes, that are wanting in the other cretaceous strata. The genera and
species must, therefore, have been spread over limited
areas; in other words, the inhabitants of the chalk
ocean had geographical limits assigned them, as is
the case with existing species.

The mode of preservation varies in the different

beds.   The shells, and stony polyparia, and radiaria of the white chalk, are generally transmuted
into carbonate of lime, with a spathose structure,
doubtless the result of high temperature, acting
under great pressure (see p. 79).   The cavities are
frequently filled with chalk, flint, or sulphuret of
iron ; in many instances they are hollow, or lined
with crystals of carbonate of lime.   The softer
zoophytes are silicified, and there is scarcely a flint
nodule in which their remains may not be traced.
The bones of animals and the coverings of crustacea
are in a friable state, and stained with sulphuret of
iron.   The teeth and scales of the fishes present a high
polish, and are coloured by a ferruginous impregnation.   Wood occurs in the state of lignite, and of
brown friable masses, which quickly decompose ;
but when enveloped in flint, the structure is well
preserved ; like the fossil wood of the tertiary,
it has evidently been drifted, and is perforated by
lithodomi ; the fissures are often filled with glittering pyrites.

In the galt, the nacreous covering of the shells is
commonly preserved, and the ammonites and nautili
of Folkstone rival in beauty the shells of the London
basin, and, like them, are subject to decomposition.
The green-sand fossils are generally silicified, and
the whetstone pits of Devonshire are celebrated for
the variety and *chalcedonic* state of the shells in
which the sandstone abounds.

The organic remains of this formation already

known, amount to many hundred species of shells, polyparia or corals, radiaria, &c. The most distinctive zoological character, is the abundance of belemnites, echinites, and ammonites: the latter are the shells of an extinct race of cephalopoda, which appears for the first time in the chalk, no traces of their remains having been discovered in the tertiary formations. My collection, consisting of many thousand fossils from the chalk formations of England and America, displays the usual species, together with many that are exceedingly rare. I will illustrate this subject with a selection of a few specimens from each class.

10. FOSSIL VEGETABLES.—The flora of the chalk, as I have already remarked, offers but little variety. Fuci, or sea-weeds, occur in some localities in great abundance. There is one species of Fucus (*Fucoides Targionii*) that abounds in the malm rock of Western Sussex, particularly at Bignor, the seat of John Hawkins, Esq., where almost every fragment of the rock is marked with its meandering forms. *Confervæ* occasionally are seen in the flints. Plants allied to *Zostera* occur in the chalk of the Isle d'Aix; drifted wood abounds in the line of junction between the galt and green sand. In the quarry of Kentish rag, of Mr. W. H. Bensted, of Maidstone, have been discovered fir-cones, leaves, large masses of perforated wood, and the stem of a plant allied to the Yucca, which Mr. Bensted has, with great liberality, placed in my collection. I reserve a

more particular description of these vegetable remains for the lecture on fossil botany.

11. CORALS, SPONGES, &c.—I have already mentioned how numerous are the softer zoophytes in
the flints. In the white chalk stony corals occur
but rarely, while the Maestricht beds contain them
in great abundance. A small *turbinolia* is not
unusual in the English chalk, and I have several
unique specimens of other genera; but the absence
of the large madrepores, and corals of that class,
is a remarkable fact, and accords with the evidence
derived from other sources, to prove that we are
examining the profound abyss of an ocean; for the
economy of the living corals fits them to live
only in waters of moderate depths. It would be
tedious to repeat to you the names which naturalists
have assigned to the fossil zoophytes of the chalk;
let it suffice to observe, that the more delicate
forms, as flustra, millepora, cellepora, &c. are very
abundant: the nature of these corals will be explained in a future lecture. There is one fossil
zoophyte, known to collectors of Sussex pebbles by
the name of *petrified sea-anemone*, from its supposed resemblance to the living actinia (Pl. II.
fig. 12). The original of this fossil was, however,
a very different creature. From an extensive suite
of specimens, I have ascertained that it was of a
subglobular form, of a tough jelly-like substance,
with a central opening or stomach, from which
numerous tubes radiated; these are exquisitely

preserved in flint. The external surface often exhibits the remains of crucial spines, similar to those
possessed by many alcyonia (Pl. I. fig. 10ᵉ).

12. RADIARIA; CRINOIDEA, &c.—The *Crinoidea,*
or lily-shaped animals, are but sparingly distributed

TAB. 30.—MARSUPITE FROM THE CHALK.

(Restored from Specimens presented by the Rev. H. Hoper, and
G. A. Coombe, Esq. of Arundel.)

in the chalk—a circumstance, as you will hereafter
find, strikingly contrasting with the zoological characters of the older secondary formations. Stems of
encrinites occur in the chalk and galt; and there is
a small species of *apiocrinite,* which is peculiar.*
The most remarkable fossil of this class is the

* Geology of the South-East of England, p. 111.

Marsupite, which I have thus named from its resemblance to a purse.

The Marsupite was a molluscous animal, of a sub-ovate form, having the mouth in the centre, and surrounded by arms or tentacula. The skeleton was composed of crustaceous, hexagonal plates, the arms, which are subdivided into numerous branches, of ossicula, or little bones; the whole was invested with a muscular tissue, or membrane. When floating, the creature could spread out the tentacula like a net, and by closing them, seize its prey and convey it to the mouth. This figure (Tab. 30) is restored from specimens which separately exhibit the parts here represented.

Asteriæ, or *star-fish*, are occasionally found in great perfection in the chalk; my friend, the Rev. Thomas Cooke, of Brighton, has discovered several remarkably fine impressions in flint, on the South Downs. The whetstone of Devonshire affords similar remains.

13. ECHINITES.—Those singular creatures, the Echini, or sea-urchins, are too well known to require minute description. Their spherical shell, or skeleton, is made up of polygonal plates, closely fitted to each other; and the surface is divided vertically, by bands like the meridians of a globe, having rows of double perforations. They are studded over with papillæ, which vary in size from mere granular points to large well-defined tubercles. To these papillæ, spines, also presenting great variety

of figure and decoration, are attached. These are
the instruments of motion, and, as on the death of
the animal, the tendons by which the spines were
attached decompose, the extreme rarity of fossil
specimens, with these processes affixed, is readily
explained. The echini, both recent and fossil,

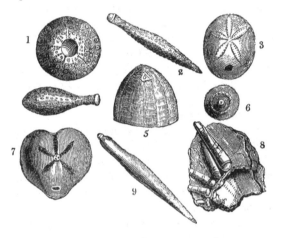

TAB. 31.—ECHINITES AND SPINES FROM THE CHALK.

Fig. 1. *Cidaris diadema.* 2, 4, 9. *Spines of Cidares.* 3. *Nucleolites.*
5. *Ananchytes cretosus.* 6. *Tubercle of a Cidaris.* 7. *Spatangus Cor-
marinum.* 8. *Spines and portion of the Shell of a Cidaris in flint.*

differ greatly in form and structure; they are arranged into numerous sub-genera, for the convenience of study, but I can only notice a few of
the usual varieties.

The helmet-shaped echinites (fig. 5) are extremely abundant, and in some localities occur in shoals, and in every gradation from the young to the adult state.   Silicious casts of echini, formed by the decomposition and removal of the shell from the flint with which they were filled, are common in gravel and on ploughed lands.   The cordiform variety (fig. 7) is very abundant, and gives rise to the heart-shaped flints of our gravel-pits.   The elliptical species (fig. 3) is common in the green sand.   The hemispherical echini are beautifully ornamented with papillæ : a small species (fig. 1) is not uncommon in the chalk and flints of Kent ; the larger varieties possess tubercles, surrounded by elegant margins (fig. 6), and are otherwise richly ornamented.   Some spines are slender and covered with asperities (fig. 2) ; others almost smooth (fig. 9), and club-shaped (fig. 4) ; it is seldom that the spines are found imbedded in contact with the shell (fig. 8).

14. SHELLS OF THE CHALK.—The bivalve shells, or conchifera of the chalk, are very numerous ; of one genus alone, *Terebratula*, above fifty species are enumerated.   Oysters, scallops, arcas, tellens, and other familiar marine shells abound, but the species differ from the recent.   With these known genera are many which, so far as our present knowledge of the inhabitants of the deep extends, are extinct.   Two or three species of *Cirrus*, or *Trochus*, are not unusual in the white chalk ; but the simple univalves are few ; and the only specimen

of a large spiral univalve with which I am acquainted is a *Dolium*, figured by Sowerby, and
belonging to Richard Weekes, Esq. of Hurstper-
point. The Maestricht beds, as I have before

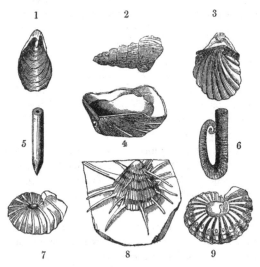

TAB. 32.—SHELLS OF THE CHALK.

Fig. 1. *Inoceramus concentricus.* 2. *Turrilites costatus.* 3. *Inoceramus
sulcatus.* 4. *Inoceramus Lamarckii.* 5. *Belemnites Listeri.* 6. *Hamites.* 7. *Ammonites Mantellii.* 8. *Plagiostoma spinosum.* 9. *Ammonites Sussexiensis.*

remarked, offer many exceptions to the usual characters of the chalk. A large *volute* (*V. Faujasii*)
is often found in the flint nodules of St. Peter's
Mountain, with *baculites*, ammonites, and other

characteristic chalk fossils. In the marl at Hamsey, near Lewes, I have discovered a few genera of simple univalves not previously known. The subglobular terebratulæ, both the common and the striated varieties, are very abundant. Another bivalve equally numerous is an elegant shell, having one valve covered with long spines (Tab. 32, fig. 8), the *Plagiostoma spinosum*,* a characteristic species of this formation. A bivalve with a fibrous structure, (*Inoceramus*, Tab. 32, fig. 4,) very brittle, and having crenulated hinges of a peculiar construction, presents numerous species; some of which are very small and delicately striated, and others two feet in diameter, and deeply furrowed. The substance of these shells closely resembles that of the recent *pinnæ;* from their fragility, fragments are very common in chalk, flint, and even in pyrites. The *Galt* contains two species of this genus, which appear to be restricted to that division of the chalk, and have been found in every locality; they are the *Inoceramus concentricus* (Tab. 32, fig. 1,) and *I. sulcatus,* (Tab. 32, fig. 3;) and a hybrid occurs in the Folkstone beds, partaking of the characters of both.

The shells of the green sand amount to many

* *Plagiostoma.*—Viscount D'Archiac informs me that the shells of this genus are true *spondyli,* and that the triangular vacancy in the lower valve is occasioned by the loss of that portion of the hinge which characterises the recent spondylus. A species of *Spherulite* (*S. Mortoni*) occurs in the Sussex chalk: hippurites, so common in the cretaceous strata of the Continent, have not been noticed.

U

hundred species: those of Devonshire are changed
into silex, jasper, and chalcedony. Dr. Fitton has
figured many of the shells from the Shanklin sand
of Kent.*

15. CEPHALOPODA, AND CHAMBERED SHELLS.
—The most peculiar and striking feature of the
population of the ocean of the chalk, as contrasted
with that of the tertiary and modern seas, is the
immense preponderance of multilocular cephalopoda.
In the tertiary, and in existing tropical seas, one
genus, the Nautilus, occurs abundantly.   The
beauty, elegant form, and remarkable internal structure of the recent shell, have rendered it in all ages
an object of admiration ; yet the nature of the
animal to which it belonged has but recently been
ascertained.   As Dr. Buckland has given a lucid
and highly interesting account both of the recent
and fossil cephalopoda,† I shall condense my remarks on this subject.

The *Sepia,* or *cuttle-fish* of our seas. is of an
oblong form, composed of a jelly-like substance,
covered with a tough skin; the mouth, which is
central, is furnished with horny mandibles, much
resembling the beak of a parrot.   The animal has
two large eyes, and eight arms, studded with rows
of little cups or suckers, which are powerful instruments both of locomotion and prehension.   The

* Dr. Fitton's Memoir on the Shanklin sands contains representations of the usual shells and zoophytes.
† Bridgewater Essay, p. 333, *et seq.*

soft body of the sepia is supported by a skeleton formed of a single bone of very extraordinary structure; in the state of dry powder it is the substance called *pounce*. The cuttle-fish has the power of secreting a dark-coloured fluid, or ink, which it ejects when pursued, and by thus rendering the water turbid, escapes from its enemies. This fluid is contained in a bag, and forms, when properly prepared, the *sepia* colour employed in the arts, and enters into the composition of Indian ink. This brief sketch of the natural history of the cuttle-fish, will enable us to understand the habits and economy of the beings whose fossil remains I am about to describe.

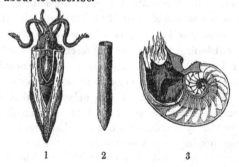

1   2   3

TAB. 33.—BELEMNITE AND NAUTILUS.

Fig. 1. *Belemno-sepia, restored, from Dr. Buckland.* 2. *Belemnites Listeri.* 3. *Section of the Shell, with the animal, of the recent Nautilus, from Professor Owen's Memoir.*

16. THE BELEMNITE AND NAUTILUS.—One of the most common fossils of the chalk is an elongated

conical stone, of a crystalline, radiated structure, and generally of a brown colour, called *belemnite·* The pits in Sussex, Kent, Norfolk, and indeed every locality of the chalk, contain these bodies; and some limestones on the Continent are almost wholly composed of them. The belemnite presents considerable variety of form, but in every species the structure consists of a spathose radiated substance, terminating in a point, (Tab. 33, fig. 2,) and having at the opposite and largest end a *conical cavity,* in which is situated a *shell* of like form, *divided into septa or chambers,* as seen in the drawing (Tab. 33, fig. 1.). Dr. Buckland has admirably explained the nature of the belemnite, and given the solution of a problem which had long been attempted in vain. The belemnite is the bone of a creature allied to the cuttle-fish, and in this representation its situation in the body of the animal, and connexion with the ink-bag, are so clearly shown, as to require no further comment. I will only add, that the belemnite sometimes occurs with the laminated, external, horny sheath,—the conical, chambered shell, —the ink-bag, —and fibro-calcareous bone, —and that the inspissated contents of the bag, the fossil sepia, has actually been made use of by one of our first artists.

NAUTILUS.—In the shell of the Nautilus we have a series of chambers pierced through the middle by a siphunculus or tube, which extends to the remotest cell. This animal is of the nature of the sepia, and occupies the outer receptacle of the shell, having a membranous tube which lines the siphuncle. The chambers are internal air-cells, and the creature has the power of filling the siphuncle only, with a fluid secreted for the purpose, and of exhausting it; and the difference thus effected in the specific gravity of the animal and its shell enables the Nautilus to sink or swim at pleasure. If, therefore, you imagine a cuttle-fish placed in the outer chamber of a Nautilus, with its arms extended, and having a tube connected with the siphunculus, but neither ink-bag or bone, these being unnecessary to an animal having the protection and mechanism of a chambered shell, you will have a tolerably correct idea of the recent and fossil Nautili. The Nautilus is essentially a ground-dwelling animal, feeding on the marine plants which grow at the bottom of the sea. " Rumphius states that it creeps with the shell above, and that by means of its tentacula it can make quick progress along the ground." *

17. The Ammonite, or Cornu Ammonis. — The fossils called Ammonites, like the Belemnites, also first appear in the secondary formations; or more properly, no memorials of their race have been found in the tertiary deposites. The Ammonite, so called from its supposed resemblance to the horns of Jupiter Ammon, is a fossil chambered shell, coiled up in the form of a disk, bearing a close analogy to

* Dr. Buckland's Essay.

the Nautilus, but differing in the situation of the
siphunculus, and in the septa by which the interior
is divided.. In the Nautilus these partitions are
entire, and their section presents a series of simple
curves, (Tab. 33, fig. 3 ;) but in the Ammonite they
possess every variety of sinuosity, and the external
surface of the casts of the Ammonites commonly
exhibits markings resembling the outlines of deeply
fringed foliage.   The shell of the Ammonite is
generally decorated with flutings, ribs, or tubercles.
I have placed before you Ammonites from the
Galt of Folkstone, in which the shell remains,—
from Watchett, with the internal nacreous coat
only,—while in this common species from Whitby
the shell is altogether wanting, the specimen being a
cast of the interior, formed of argillaceous iron-stone,
a state in which these fossils are frequently found.
In these examples from Mr. Bensted's quarry, the
shells and partitions of the separate chambers having
decomposed, casts have been formed, which fit into
each other, and admit of being put together, so as to
show the distinct shape of the Ammonite.   Nautili
also occur in this state ; and I have a specimen for
which I am indebted to Miss Pearson, of Clapham,
in which the series is complete from the commencement to the outer cell.   Nearly three hundred kinds
of Ammonites are known in the secondary formations, certain species being restricted to particular deposits.   Thus, for example, the chalk marl of
Sussex abounds in two kinds of Ammonites, (Tab.

32, fig. 7, 9,) which either are very rare, or do not occur in the white chalk above, or in the blue clay below; and in every part of England, and on the Continent, these species are found in the same geological position. But I must again refer you to Dr. Buckland's Bridgewater Essay for some most important and highly interesting information on these subjects. I will add but one remark; the membranous tube of the air-cells sometimes occurs in a fossil state, as may be seen in this Ammonite from the chalk marl of Lewes, which retains a large portion of the siphuncle; the black substance of these tubes has been analysed, and found to consist of animal membrane, permeated by carbonate of lime. Dr. Prout explains the black colour to have originated from decomposition; the oxygen and hydrogen of the animal membrane escaping, and carbon being evolved, as happens when vegetable matter is converted into coal, under the process of mineralization. The lime has taken the place of the oxygen and hydrogen, which existed in the pipe before decomposition.*

Ammonites vary in size from a few lines to twelve or fourteen feet in circumference; at low water on those parts of the Sussex coast where the chalk forms the base of the shore, enormous Ammonites are often seen imbedded. In some limestone districts, the marble is almost wholly composed of Ammonites, as in this polished slab from Somerset

* Bridgewater Essay, p. 352.

shire, which exposes the most beautiful and varied
sections of the shells.

Baculités, Turrilites, Hamites, and other genera of
multilocular shells, abound in the Chalk marl, Galt,
and Shanklin sand. The Turrilite (Tab. 32, fig. 2,)
may be described as an Ammonite twisted in a spiral,
instead of a discoidal form: and the Hamite,
(Tab. 32, fig. 6,) as a similar structure in the shape
of a hook, coiled up at the smaller extremity. These
shells sometimes attain a large size; the Turrilite
before you, which is the finest example known, would
if perfect exceed two feet in length; it possesses traces of the siphuncle. Hamites of gigantic
proportions have been found in the Shanklin sand of
Kent, by Mr. Hills, the intelligent and indefatigable
curator of the Chichester Museum. The first specimens of Turrilites, Hamites, and Scaphites, discovered
in the British strata, were, in my early researches,
found in Hamsey marl-pits, near Lewes. The scaphite
is of a boat-like form; but I must forbear entering
on its description, as well as on that of many other
interesting kinds of multilocular shells, hundreds
of which are microscopic, and sometimes form
entire masses of chalk, as Mr. Lonsdale's researches
have shown.

18. SPIROLINITES.—There is however one genus
which I cannot pass over without remark; it is
called, from the disposition of its chambers, the
Spirolinite, and resembles the common *Nautilus
spirula,* the *crozier-shell* of collectors, except that

the coils, which in the recent shell are separate, are
in close apposition in this fossil. Several species of

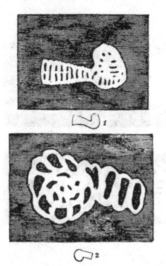

TAB. 34.—SPIROLINITES IN FLINT. DISCOVERED BY THE
MARQUIS OF NORTHAMPTON.

Fig. 1. *Spirolinites Lyellii.* *Fig.* 2. *S. Comptoni.*

this genus, previously unknown in the chalk, were
discovered by the Marquis of Northampton some
years since, in the pebbles on the Brighton shores;
and in flints from the chalk at Kemp-town, and other
places: these minute but most interesting objects
having escaped the notice of less accurate and intelligent observers. My son has also collected

several specimens from the flints around Chichester.
I have great pleasure in laying before you enlarged

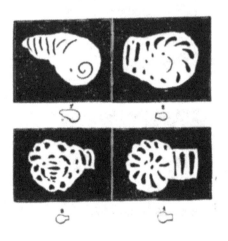

TAB. 35.—*Fig.* 1. *Spirolinites Stokesii.* 2. *S. Mantellii.* 3. *S. Lyellii.*
4. *S. Comptoni.*

drawings, by his Lordship, of four species ; the
smaller figures indicate the size of the originals.*

* Note on the Sussex Spirolinites, by the Marquis of Northampton :—

" I willingly comply with your desire to communicate a short
note on the Sussex Spirolinites, one species of which you have
been pleased to distinguish by my name. I have found these fossils
in flint at Brighton, Kemp-town, Rottingdean, Lewes, Hastings,
Steyning, Chichester, West Stoke, and in the Isle of Wight ;

19. INFUSORIA IN FLINT.—Among the almost innumerable diversity of forms in which those atoms of animal existence, the *Infusoria* (so called because many species abound in vegetable infusions), appear, many species like the Cypris, of which we have already spoken, possess shields or coverings, some of which are ferruginous, others calcareous, and many silicious; the ferruginous film seen on the surface of stagnant water is made up of Infusoria. The Infusoria

and one specimen in France. I have discovered about two hundred of these minute chambered shells in flint, but only two in chalk. Some of the microscopic bodies extricated from the chalk by Mr. Lonsdale, I am inclined to think are Spirolinites, but others are *foraminifera.* I have seen, as I believe, minute nummulites in the Sussex flint. The Spirolinites which I have collected constitute four distinct species. 1. The one to which you have been pleased to give the appellation of *S. Comptoni.* 2. *Spirolinites Lyellii,* distinguished by the horizontal chambers, one above the other, in the coiled portion. 3. *S. Stokesii,* which I name after our friend Charles Stokes, Esq.; and this name has the further advantage of pointing out the locality, West Stoke, near Chichester, from which I obtained this unique specimen, and where other Spirolinites abound. The fourth you must allow me to designate *S. Mantellii.* The distinctive characters of these species are too obvious to require detailed description. The transverse chambers in *S. Lyellii* (Tab. 35, fig. 3,) are a striking peculiarity of structure; in the specimen from Hastings, there appear indications of a siphuncle in the straight prolongation (Tab. 34, fig. 1). I am inclined to believe that there are other species in my collection, but the irregularity in the fractured sections of these minute chambered shells, renders it difficult to arrive at accurate conclusions on this point.

" *Castle Ashby.* "   " NORTHAMPTON."

belong to many distinct families, some having a complex organization, with a nervous, muscular and circulating system, and digestive organs highly developed. As I shall revert to this subject in another lecture, I now only wish to call your attention to the remarkable fact, that the silicious cases, or skeletons, of this class of beings, have been discovered in a fossil state; and that some deposites, for instance the tripoli of Bilin in Bohemia, consist almost entirely of the silicious remains of Infusoria, of a species so minute, that a cubic inch of stone, weighing 220 grains, contains upwards of 41 millions of these skeletons.* The distinguished naturalist, Ehrenberg, to whom we are indebted for this wonderful discovery, has also detected the remains of these animalculæ in chalk-flints, semi-opal, and other silicious substances; and the Rev. I. B. Reade, of Peckham, has observed in the flints of Surrey shields of *Gaillonella*, a form of infusorial animal well known to microscopic observers. I shall hereafter place before you representations of some of these objects, with which Mr. Reade has kindly favoured me; it is sufficient for the present to have stated the fact, that *entire masses of flint are composed of the fossilized remains of beings,* as wonderful in their structure and organization, as any of the colossal forms of animal existence.

* See a translation of Ehrenberg's Observations on these Discoveries, in Taylor's Scientific Memoirs, Part III.

20. CRUSTACEA.—Species of several genera of crustacea have been extricated from the Sussex chalk; in some examples I have succeeded in removing the surrounding stone, and exposing the filiform antennæ, the abdominal segments, and the tails of *Astacidæ*. In the Galt, the crustacea hitherto discovered belong to very small species. I have obtained from Ringmer, near Lewes, specimens which, in the opinion of Dr. Leach, are extinct forms, related to Indian genera. In the Speeton clay of Yorkshire, Professor Phillips has discovered several beautiful species of Astacus. The Shanklin sands of Kent and Dorsetshire have yielded a few crustacean remains. In the limestone of St. Peter's Mountain, claws of a small kind of crab are frequently discovered, but no other vestiges of the animal. Faujas St. Fond and Latreille have very ingeniously explained this fact, by showing that the claws belonged to a parasitical species, which, like the common hermit-crab of our seas, had the body covered by a delicate membrane, the claws alone having a shelly case; hence the latter would be found in a fossil state, while of the other parts of the animal no traces would remain.

21. FISHES OF THE CHALK—SHARKS.— The fossil fishes of the chalk were known only by the teeth, with which almost every quarry abounds, until my researches in the chalk-pits around Lewes brought to light the extraordinary specimens before you, and showed how such delicate remains could

be developed. The teeth are for the most part
referrible to fishes allied to the shark; a family
which in the ancient, as in the modern seas, appears
to have been confined by no geographical limits.
Professor Agassiz, by whose genius and perseverance
this department of Palæontology has been so successfully elucidated, has proposed a classification of
fishes, founded upon the peculiar structure of the
scales — an arrangement of great utility to the
geologist, since the mutilated state in which ichthyolites frequently occur, had rendered futile the
attempt to place them in existing orders and
genera.*

The teeth are very abundant, particularly those belonging to the Lamna (Tab. 36. fig. 2, 4); they possess
a high polish, are in an excellent state of preservation,
and occur but rarely in the flints. They are always
single, arising from the cartilaginous nature of the
jaws of the original. These examples of the recent
shark show the number and variety of the teeth in
an individual; by the decomposition of the jaw the
teeth would be separated and drifted by the water,
and therefore seldom, in a fossil state, exhibit any
traces of their original position. It may, however,
happen, that jaws with teeth will hereafter be
discovered; as vertebræ, fin-bones, and even the
shagreen-skin of sharks, are preserved in some
specimens in my collection. The broad rugous

* See Dr. Buckland's Essay, p. 268.

teeth (fig. 6) are sometimes found in groups of twenty or thirty; they belong to a fish allied to the shark, in which the mouth was covered with these

TAB. 36.—TEETH OF FISHES ALLIED TO THE SHARK, FROM THE SUSSEX CHALK.

Fig. 1. *Notidanus microdon.* 2. *Lamna Mantellii.* 3. *Galeus pristo-dontus.* 4. *Lamna appendiculata.* 5. *Ptychodus altior.* 6. *Ptychodus decurrens.*

bony processes, like a tessellated pavement. The spines, fin-bones, or rays, with which fishes are furnished, also occur in the chalk; and I have one splendid specimen, in which even the tendinous expansion of the muscle that moved the fin-bone is preserved.* Some of those in my cabinet belong to the same genus as the dog-fish of our coast (*Spinax Acanthias*), which has a curved spine in front of the dorsal fin ; I place before you a recent and fossil spine, to show their analogy. The mandible, or jaw-bone, of a very curious fish, the *Chimera*, was one of my earliest discoveries in Hamsey marl-pit, and I have since found examples in the chalk of Lewes ; other species have been discovered in the green sand of Kent, by Mr. Bensted, and in Kimmeridge clay, by Sir Philip Egerton.†

Remains of large fishes, belonging to that division called by Agassiz, *Sauroid*, from their combining in their structure certain characters of reptiles, have been found in the chalk and green sand of Sussex and Kent. They consist of large, conical, striated teeth, bearing a resemblance to those of crocodiles, with which they were formerly confounded. Mr. Bensted has several from Maidstone, and Mr. Rose, from the galt, near Cambridge.

22. Fossil Salmon, or Smelt.—But the most remarkable ichthyolites of the chalk, are the fishes belonging to the salmon family (*Salmonidæ*), and closely related to the smelt (*Osmerus*). Many years since, I succeeded in extricating from the chalk the extraordinary specimen before you, of

* Fossils of the South Downs, Tab. xxxix.

† The nature of these curious relics remained unknown, till Dr. Buckland ascertained their analogy to the recent Chimera.

which I published a lithograph, and described the fish under the name of Salmo Lewesiensis.*

The length of the fish is nine inches, and it stands nearly six inches in relief; the back is still attached to the chalk, and the dorsal fin exposed. There are other examples of the same species in my cabinet, which are almost equally perfect. These ichthyolites were obtained from the quarries in the immediate vicinity of Lewes, during my residence in that town. It is clear that the chalk must have surrounded the fishes while they were alive, and in actual progression, and by suddenly consolidating, have preserved their forms unaltered; for the body is round and uncompressed, the mouth open, and the fins and gills are expanded. Even to those whose curiosity has not previously been awakened to geological inquiries, the examination of these petrified inhabitants of the ocean cannot fail to excite deep interest, and I have seen the man of fashion, as well as the philosopher, gaze in mute astonishment on these "relics of a former world."

23. MACROPOMA AND OTHER FISHES.—I have already mentioned that the capsule of the eye remains in many specimens; this is particularly the case with those fishes (Beryx) which have some resemblance to the Dory (Tabs. 40, 41); and in a sauroid fish, the membranes of the stomach are

---

* Geology of the South-East of England, p. 135. A beautiful lithograph of this specimen, by Mr. Pollard, forms the embellishment of the Catalogue of the Mantellian Museum.

invariably preserved. This fish, named *Macropoma*, (Tab. 38) by M. Agassiz, is, independently of the fact just stated, extremely remarkable in its organization. The operculum of the gills is very large, and the scales are studded with hollow tubes. In many recent fishes, there is a row of tubular scales, forming what is called the lateral line, through which flows a fluid that lubricates the surface of the body; in the *Macropoma*, every scale appears to have possessed such a mechanism.

Many of the most interesting chalk ichthyolites in my museum are figured, and occupy more than twenty folio plates, in the last *livraisons* of the splendid work of Agassiz, " Recherches sur les Poissons Fossiles." I will now place before you restored figures of seven species; for comparative anatomy enables us not only to reanimate the colossal mammalia, and the palæotheria, but also to restore, with all the lineaments of life, the fishes which lived and died in the seas of the ancient world. These restorations have been drawn with great care by an eminent artist, M. Dinkel, of Munich.

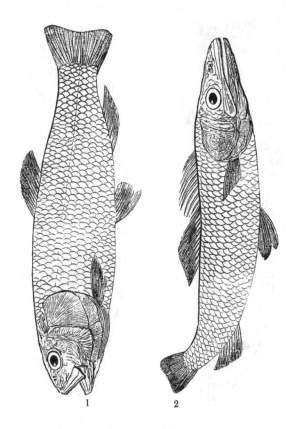

TAB. 37.—Fig. 1. OSMEROIDES MANTELLII. Length 12 inches. From
Lewes Chalk-pits.
Fig. 2. ACROGNATHUS BOOPS. Natural size. *Unique.* From Lewes.

x 2

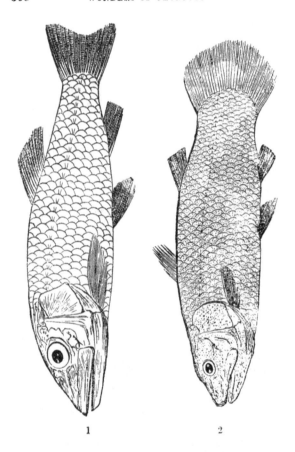

1   2

TAB. 38.—Fig. 1. AULOLEPIS TYPUS.—Length 6 inches. *Unique.* From Clayton Chalk-pit. Fig. 2. MACROPOMA MANTELLII.—Length 24 inches. From the Chalk quarries near Lewes.

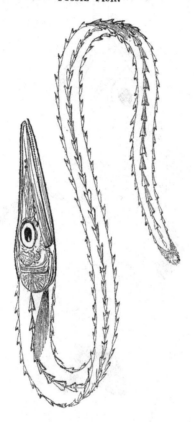

Tab. 39.—Dercetis elongatus.—Length 16 inches. From Lewes.

This species occurs abundantly in the chalk at Preston, near Brighton. The outline represents the skeleton, from a specimen in my museum; the only instance in which the skull remains.

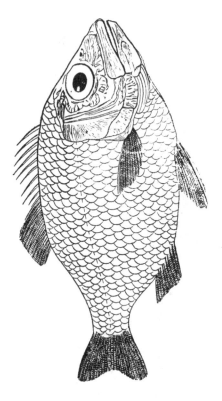

Tab. 40. — Beryx radians. Length 7 inches.—From the chalk-marl, near Lewes.

This species is generally found in the chalk-marl : specimens have been collected at Clayton, Steyning, and Arundel.

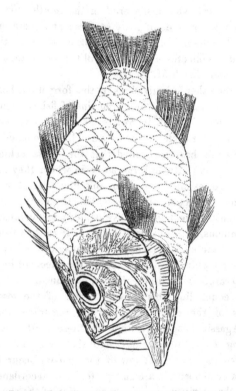

Tab. 41.—Beryx Lewesiensis. Length 12 inches.—Lewes chalk quarries.

This is the most abundant of the Sussex ichthyolites; and is called *Johny Dory* by the workmen. Detached scales are very frequent in all the pits of the South Downs, and also in those of Kent and Surrey.

The fossil fishes discovered in the South Downs amount to upwards of forty species; and there are several undescribed from the chalk of Kent, in the splendid collections of Ichthyolites of Viscount Cole, and Sir P. M. Egerton.

In the other sub-divisions of this formation, both in England and elsewhere, remains of fishes occur. The slates of Glaris, in Switzerland, have long been celebrated for their ichthyolites, and by these remains M. Agassiz has ascertained that the strata belong to the chalk; although the stone in which they are imbedded, as may be seen in this fine suite of specimens (collected and presented to me by the distinguished geologists above named), is a compact bituminous slate, scarcely to be distinguished from some of the most ancient of the transition series; a character which is attributable to the effects of high temperature, as will hereafter be explained.

In concluding this cursory review of the fossil fishes of the chalk, it must be remarked that M. Agassiz has shown that all these ichthyolites belong to extinct forms; and that none of the species, and even but few of the genera, occur in the later deposites; a result perfectly in accordance with that derived from the examination of the shells and mollusca.*

24. REPTILES.—The remains of reptiles hitherto observed are but few, the most important is the *Mosæsaurus Hoffmanni*, of which I spoke when

* Appendix I.

describing the Maestricht deposites. The occurrence of the vertebræ of the same genus, if not species, in the Lewes chalk, and of the teeth and bones in the equivalents of this formation in North America, are facts of great interest. Through the kindness of Mr. Charlesworth, I have inspected portions of a large jaw with teeth from the Norfolk chalk, which bear a resemblance to those of the mosæsaurus; but the symmetrical, conical form of the teeth, and other characters, show that they belong either to an unknown reptile or to a sauroid fish. Bones of turtles are found in the white chalk of Sussex, and abundantly in the limestone of St. Peter's Mountain, and in the slate of Glaris; they belong to marine species. I have a mandible of a turtle from the Lewes chalk, figured by Dr. Buckland, (Essay, Pl. XLIV. fig. 3;) and a femur from Kent, discovered by Mr. Bensted. Teeth of crocodiles in the chalk of Meudon are mentioned by Cuvier; and very recently a specimen containing the vertebral column, ribs, and pelvis of a small lizard, in a beautiful state of preservation, has been found in a chalk-pit near Chatham, and is in the possession of Sir P. M. Egerton.

25. REVIEW OF THE CHARACTERS OF THE CHALK FORMATION.—The characters of the chalk formation, as shown by these investigations, are those of a vast oceanic basin, filled with the debris thrown down by its waters, and enveloping the remains of its inhabitants; arenaceous beds

prevailing in the lowermost—argillaceous in the middle—and cretaceous in the upper division of the series. Intrusions of thermal streams appear to have been abundant at certain periods; and throughout the entire epoch of its formation, the proofs are incontrovertible that its waters swarmed with living beings of the various orders of marine existence; the species being all, or almost all, extinct. The presence of fuci shows that it possessed a marine vegetation; and drifted wood, fir-cones, stems, and leaves, that its shores were bounded by dry land, of which the fossil reptiles also afford additional confirmation.

26. GEOLOGY OF THE SOUTH-EAST OF ENG-LAND.—From this survey of the vast marine formation of the chalk, we turn to the remarkable fluviatile deposites, of which the basin of that ocean, in the south-east of England, was composed; in other countries, as I shall again have occasion to remark, that basin was formed of the oolite and other marine strata. It will now be necessary to offer a few observations on the geology of the district in which the beds of the Wealden are so largely developed.

The deposites of the south-east of England constitute three principal groups. The *first* consists of the *tertiary* beds of sands, clays, and gravel, described in the previous lecture, which occupy depressions of the chalk. The *second* is the *chalk*, which forms the most striking feature in the physical geography of the country, constituting the South Downs, which from the bold promontory of Beachy-head, traverse the county of Sussex from east to west, and pass by Hampshire into Surrey. From Godalming the chalk extends by Godstone into Kent, where the range is called the *North Downs;* and terminates in the cliffs of Dover. The *third* group is spread over the area between the North and South Downs; the most elevated masses forming a range called the *Forest-ridge*, which traverses the district in a direction nearly east and west, and is composed of alternations of sandstone, sands, shales, and clays, with a deep valley on each flank, called the *Weald;* hence the geological designation of the whole series. From the central ridge of the Wealden, which varies in height from 400 to 800 feet, the strata diverge on each side towards the Downs, forming an *anticlinal axis*, and finally disappear beneath the lowermost beds of the chalk. (*Vide* the Section, Pl. V. fig. 1.) There are conclusive proofs that the Wealden strata were entirely covered by the chalk, and that their present position and appearance are attributable to changes which have taken place subsequently to the cretaceous epoch; the Wealden having been lifted up and forced through the chalk, and thus effected the partial destruction of that formation.*

27. GEOLOGICAL PHENOMENA BETWEEN LONDON AND BRIGHTON. — The direct roads from

* See Geology of the South-East of England, Chapter XI.

London to Brighton pass over the whole series of
deposites comprised in the above sketch, as well as
over those described in the first lecture.   Proceeding from the Thames, the observer successively
traverses the modern silt of the river—the ancient
alluvium, containing remains of elephants and other
large mammalia—and if he proceed by Reigate, his
road, at Clapham and Tooting, lies over beds of clay
and gravel, which are part of the ancient shores
of the London basin.   At Sutton he ascends the
chalk hills of Surrey, and travels along elevated
masses of the ancient ocean-bed just described.
Arriving at the precipitous southern escarpment
of the North Downs, a magnificent landscape, displaying the physical structure of the Weald, and
its varied and picturesque scenery, suddenly bursts
upon his view.   At his feet lies the deep valley of
Galt in which Reigate is situated, and immediately
beyond the town is the elevated ridge of Shanklin
sand, which stretching towards the west, attains at
Leith-hill an altitude of one thousand feet; and
to the east forms a line of sand-hills, by Godstone
and Sevenoaks, through Kent, to the sea-shore.
The Forest-ridge occupies the middle region, extending westward towards Horsham, and eastward
to Crowborough-hill, its greatest altitude, and from
thence to Hastings, having on each flank the
Wealds of Kent and Sussex; while in the remote
distance, the unbroken, undulating summits of the
South Downs appear like masses of grey clouds
on the verge of the horizon.

Pursuing his route, the observer passes through Reigate, over the *Galt,* (see Pl. V. Section I. 2,) and *Shanklin sands* of Reigate-hill, (3,) and arrives at the commencement of the Wealden. The *weald clay,* (4,) containing limestone with fresh-water shells, appears at Horley common; and while in the commencement of his journey the roads were made of broken chalk-flints, and at Reigate of cherty sandstone, the material here employed is the bluish-grey calcareous rock of the Weald. At Crawley (5,) sand and sandstone appear, and the road is composed of grit and stone, containing fluviatile shells, bones, and plants. Crossing *Tilgate Forest,* and Handcross, over a succession of elevated ridges of sandstone, and of clay valleys, produced by alternations in the strata, he descends from the sandstone ridge at Bolney, and again journeys over a district of weald clay with fresh-water lime-stone (4, *on the left*). *Shanklin sand,* like that of Reigate, reappears at Hickstead (3, *on the left*), succeeded by a tract of *Galt* (2, *on the left*) ; and finally, entering a vale of chalk-marl, he reaches a defile in the South Downs, through which the road winds its way to Brighton; the traveller having in the course of his journey passed from one chalk range to the other, and traversed the ancient delta of the Wealden.*

28. The Wealden. — The tertiary basin of'

* The reader will be able to follow this route by referring to the Section, Pl. V. fig. 1.

London has afforded us an illustration of the
process by which materials are accumulated, and
organic remains imbedded, in an inland sea, — that
of Paris, of marine and fresh-water sediments, deposited in a gulf open to the sea on one side, and
fed by rivers and thermal springs on the other,—
the lacustrine formations of Auvergne, of the
gradual precipitation of strata in the tranquil waters
of lakes,—the chalk, of the operations which have
taken place in the profound abyss of an ocean,—
while the series of deposites to which the term *Wealden* is applied, presents the most striking example of
an ancient fluviatile formation hitherto discovered.
Yet strange as it may appear, although the Wealden strata are spread over the whole area between
the North and South Downs, a tract of country
traversed daily by hundreds of intelligent persons
from the metropolis, their peculiar characters were
unknown fifteen years ago ;* the entire group

* " Until the appearance of Dr. Mantell's works on the
Geology of Sussex, the peculiar relations of the sandstones
and clays of the interior of Kent, Sussex, and Hampshire, were
entirely misunderstood. No one supposed that these immense
strata were altogether of a peculiar type, and interpolated
amid the rest of the marine formations, as a local fresh-water
deposit, of which only very faint traces can be perceived in
other parts of England."—*Professor Phillips, Ency. Met.* p.631.
Art. *Geology.*

" It was not until the appearance of Dr. Mantell's Illustrations of the Geology of Sussex, in 1822, that the full value
of the evidence which this district affords was made to appear.
In that excellent work the author clearly showed that the

being supposed by geologists to belong to a series of marine clays and sands below the chalk.

Before entering upon the description of the strata, I would remind you of what has been stated in a previous lecture, of the effects of rivers, and the nature of modern fluviatile deposites (pp. 34, 35). We found the deltas of rivers to consist of clay (or indurated mud), alternating with beds of sand and sandstone (or consolidated sand), and containing leaves, branches, and trunks of trees, fresh-water shells, works of art, bones of man, and of land animals, more or less rolled,—with boulders formed of fragments of rocks, transported by torrents from the hills, or washed out of the banks by the streams. Let us now suppose that by agencies already explained, a river has disappeared, that the sea also has changed its place, and the bed and delta of the river become dry land; that towns and villages have been built upon the now consolidated delta, and that its surface is clothed with woods and forests, or under cultivation. If sections of the strata were exposed, either by natural or artificial means, and the bones of man or animals, works of art, and remains of plants and shells, were visible in the clay or sand-stone, such appear-

extraordinary remains which he had discovered in the beds of Tilgate Forest must have originated in a lake or estuary, and have been the produce of a climate much warmer than that which is now enjoyed in England."—*Dr. Fitton's Geology Hastings*, p. 14.

ances would excite in us no surprise, because we are acquainted with the processes by which such accumulations of water-worn materials are formed. Should an inhabitant of the new country express his wonder how brittle shells, delicate leaves, and bones of animals had become imbedded in the solid rock, and if when we stated the manner in which those changes had been effected, he should not only refuse his assent, but insist that the shells, leaves, and bones, were merely accidental forms of the stone, should we not feel astonished at his ignorance and prejudice? yet not a century since, and such an opinion almost universally prevailed, and is even still entertained by many!* And farther, if our assumed personage admitted that the remains in question were fossil animals and vegetables, but asserted that they had been entombed in the strata by a deluge which had softened the crust of the earth, and engulfed in the sediment of its waters the remains of animated nature, — should we not reply, that as such a catastrophe must inevitably have mingled together all kinds of materials, and

* " At Hawkhurst, in the Weald of Kent, these stones (Sussex marble) abound. They consist of several laminæ, between which grow shells, or rather half-shells, having the appearance of periwinkles of different magnitudes, according to the time of their growth. These stones naturally grow in the earth, and the shells upon them, and are another certain proof that shells are generated in the earth, as well as in the sea, and that there is no necessary connexion between a shell and an animal."— *Natural History of England*, p. 193, vol. i. 1759.

the remains of animals and vegetables, whether of
the land, the rivers, or the seas—the regular stratification of the delta, and the exclusive occurrence of
land and fresh-water plants and animals, were fatal
to such a supposition, and afforded conclusive evidence of the correctness of our explanation of the
phenomena?—it is by such a train of reasoning
that the fluviatile nature of the Wealden has been
determined.

29. WEALDEN OF THE SUSSEX COAST. — From
the distribution of the Wealden over the south-east
of England, instructive sections have been formed by
the action of the sea along the coast, between Beachy
Head and Dover. From the stupendous cliffs of
Beachy Head the chalk extends towards Southbourn,
where beds of galt, glauconite, marl, and Shanklin-
sand successively emerge, forming the base of the
shore, and abounding in their characteristic marine
fossils. Passing over Pevensey Levels, the boundary
of which, on the sea-side, is obscured by modern
shingle, we arrive at Bexhill and Bulverhithe, and
find the cliffs composed of finely laminated sandstone and clays; and those of St. Leonard, of
similar strata, more extensively developed: sands
and clays separated into very thin laminæ, alternate with conglomerates, indurated sand-rock, and
a fine sandstone, called *grit*, of great compact-
ness. At Hastings, strata of sand and clay, with
interspersions of lignite, laminated shale, grit,
and sandstone, constitute a long line of high

cliffs.* The general resemblance of these strata to fluviatile accumulations is most striking; the laminated structure of the clay and shales, the constant intermixture of minute portions of lignite, the absence of pebbles and shingle, and the alternations of mud and sand, are lithological characters constantly observable in river deposites. The nature of the organic remains in which the strata abound will be considered hereafter. In the interior of the country the quarries opened along the ridges, formed by the compact grit, afford various sections; and the valleys, eroded by the streams, expose the shales and laminated clays.

30. POUNCEFORD.—Pounceford, on the estate of Lord Ashburnham, on the road to Burwash, in Sussex, presents several highly interesting sections of the argillaceous beds and limestones. Descending through a defile in the sandstones, we arrive at the bottom of a glen, along which a rapid stream, that bursts out from between the clay-partings, rushes to a distant and lower valley. On each side openings are made, to arrive at a greyish blue limestone, abounding in shells, which is employed on the roads, and converted into lime for agricultural purposes. When the stone lies deep, shafts are sunk from the surface, and after the extraction of the limestone, are deserted and filled up.

* See an excellent little volume on these cliffs, "*A Guide to the Geology of Hastings*," by W. H. Fitton, Esq. M.D. F.R.S. &c.

This spot is highly interesting and picturesque;
incrusting springs, issuing from the limestone beds,
deposit tufa on the mosses, equiseta, and land-shells:
thousands of fossil shells are strewn over the clay
and shale; and stems of plants, scales of fishes,
and other remains, are seen imbedded in the
stone; while the banks, where newly exposed, exhibit countless laminæ and alternations of shale,
clay, and layers composed of testaceous remains.* In a visit to this place with Mr. Lyell,
several new species of fossil shells were found in
the bed of the stream, having been washed out of
the banks of clay; and we collected teeth of crocodiles, bones of fresh-water turtles, and other reptiles. A spiral fresh-water shell (Tab. 46, fig. 2,)
was abundant in the clay; and a muscle, (named
Mytilus Lyellii, to commemorate our excursion,
Tab. 46, fig. 8,) also a fluviatile species, was found
by Mr. Lyell, in a mass of shale that had fallen
into the stream.

As the *grit*, or calciferous sandstone, of the
Wealden, forms an excellent road-material, the
quarries along the principal lines leading from the
metropolis to the south-eastern coast, are very
numerous; and those spread over the area of Tilgate and St. Leonard's Forests have been extensively
worked since the increased communication between London and Brighton. This district may be

---

* See Geology of the South-East of England, p. 22.

described as bounded on the west by the London
roads leading through Horsham, and on the east
by those which pass by Linfield or Cuckfield;
the Crawley road, as previously mentioned, traversing Tilgate Forest. These localities, particularly that of Tilgate, have acquired much celebrity
for their organic remains, from having been the
principal sources whence the specimens figured
in my first work on the "Fossils of Tilgate
Forest,"* were derived; but every quarry throughout the Forest-range, from Loxwood in western
Sussex, to Hastings, will be found to yield the
peculiar remains of the Wealden, more or less
abundantly.

31. SUBDIVISION AND EXTENT OF THE
WEALDEN.—The Wealden may be divided into
several groups, each characterized by the nature of
the strata, and the prevalence of certain kinds of
fossils; but throughout the whole series, the fluviatile nature of the formation is maintained: in the
lowermost part of the series only, are there any
intrusions of a marine or estuary nature. Although
it is not within the scope of these lectures to enter
upon minute details of stratification, it will be
necessary, for the elucidation of the subject, to point
out the principal subdivisions of this extensive system of deposites.†

---

* See Fossils of Tilgate Forest, p. 51.
† Geology of the South-East of England, p. 182.

1. WEALD CLAY (*the uppermost or latest deposit*). Stiff blue clay, with septaria, argillaceous ironstone, and beds of shelly limestone, called *Sussex or Petworth Marble*. (See the Section, Pl. V. fig. 1. 4.)
2. HASTINGS BEDS. Sand and sandstones, with calciferous grit, or *Tilgate-stone*, alternating with clays and limestones.
3. THE ASHBURNHAM BEDS. Clays, shales, and bluish-grey limestones, and sandstones.
4. THE PURBECK BEDS. Clays, sandstones, and shelly limestones, called *Purbeck Marble*. Limestone, with layers of *vegetable mould*, and remains of trees in a vertical position—the petrified *Forest of Portland*.

Such is the assemblage of deposites which the term Wealden, first employed in this acceptation by my friend Mr. Martin,* is intended to denote. Clays or argillaceous sediments, with limestone almost wholly composed of fresh-water snail-shells, occupy the uppermost place in the series ; sand and sandstones, with shales and lignite, prevail in the middle ; while in the lowermost, argillaceous beds, with shelly marbles or limestones, again appear ; and, buried beneath the whole, is a petrified forest, in which the trees are still standing, and the vegetable mould undisturbed ! The upper clay-beds and marble form the deep valleys or Wealds of Kent and Sussex — the middle series, the forest-ridge. The Purbeck are obscurely seen in some of the deepest valleys of eastern Sussex, but emerge on the Dorsetshire coast and form the island or peninsula whose name they bear, and surmount the

---

* Martin's Geology of Western Sussex.

northern brow of the Isle of Portland. At the back
of the Isle of Wight, the Wealden beds appear
beneath the Shanklin sands; and their characteristic
fossils are continually washed up on the shore at
Brook-point. Dr. Fitton* has traced the Wealden
beds, or rather the lowermost division, the Purbeck,
in the vale of Wardour, which is a valley of denudation, in the south of Wiltshire, like that of the
South-east of England, on a small scale. In this
valley the various members of the chalk occur in
their regular order of superposition, resting on clay
and Purbeck limestone, and having Portland stone
beneath.† In France, on the coast of the Lower
Boulonnais, and in the valley of Bray near Beauvais, strata of a like character are observable; the
Sussex marble (*lumachelle-à-paludines*) and a fern
peculiar to the Wealden, have been discovered by
M. Graves of Beauvais, to whom I am indebted for
specimens. It is probable that the Wealden may
have extended over a still larger area, for the same
fossil plant (*lonchopteris Mantellii*) has been found
in strata beneath the green-sand, in Sweden, by
Professor Nillson, who also informed me that
many of the rocks and fossil plants from Tilgate
Forest were analogous to specimens he had observed in the little island of Bornholm, off the
Danish coast. Without relying upon these obser-

* Consult Dr. Fitton's Memoir "On the Beds below the
Chalk;" Geological Transactions, 1837.
    † Ibid. p. 424.

vations, the Wealden may be considered as covering an area 200 miles in length, from west to east, and 220 miles from north-west to south-east; an extent but little exceeding the delta of the Ganges, or the Mississippi, and not equal to that of the Quorra, which forms a surface of 25,000 square miles, being equal to the half of England; the total thickness of the deposites averages about 2000 feet.*

32. QUARRIES OF TILGATE FOREST.—The quarries of Tilgate Forest, where the calciferous grit is worked, generally present the following series of deposites :—

1. *Uppermost.* Loam or clay—from one, to five or six feet in depth. Destitute of fossils.
2. Sandstone—friable, of various shades of fawn, yellow, and ferruginous colour; in laminæ, or thin layers, occasionally containing organic remains and pebbles,—eight feet thick.
3. *Calciferous Grit,* or *Tilgate Stone*—a very fine sandstone, formed of sand cemented together by calcareous spar; it occurs in large masses of a concretional form, imbedded in beds of sand. This grit has evidently been formed by the percolation of water charged with calcareous matter, into loose sand; it abounds in bones and teeth of reptiles; stems and leaves of plants; shells, &c.
4. Sandstone, with concretionary masses of grit and conglomerates formed of rolled pebbles of sandstone, jasper, quartz, indurated clay, bones, and teeth of reptiles and fishes; rolled masses of the grit and sandstone are found in this conglomerate; contains organic remains which are generally much water-worn.
5. Blue clay and marl—depth unknown.

* Dr. Fitton.

Such is the general aspect of the quarries around Bolney, Cuckfield, Linfield, &c. Near Horsham the fawn-coloured sandstone is of a more compact character, and possesses a slaty structure. The thin slates are used for roofing, and the thicker beds afford good paving-stone; their surfaces are sometimes deeply furrowed by ripple-marks—an appearance on which I will offer a few observations.

33. RIPPLE-MARKS ON SANDSTONE.—The furrowed sandstone and grit which are used for paving in Horsham, Crawley, and other towns and villages on the Forest-ridge and Tilgate Forest, must have attracted the attention of most persons who have travelled from Brighton to London. The surface of these slabs is similar to what may be observed on the sand along the sea-shore at low water, when the ripple from the receding waves has been well marked; and the appearance has arisen from a similar cause. (See p. 30.) In many instances the stone is so rough as to be employed in stable-yards, where an uneven surface is required to prevent the feet of animals from slipping in passing over. It sometimes happens that when a large area of a quarry is cleared from the soil which covers it, a most interesting appearance is presented, the whole surface being rippled over like the strand on the sea-shore; and the spectator is struck with the conviction that he is standing on the sands of some ancient delta or estuary, which are now turned into

stone. Sometimes the furrows are deep, showing
that the water was much agitated, and the ripple
strong ; in other instances the undulations are
gentle, and intersected by cross ripples, proving a
change in the direction of the waves. Some slabs
are covered by slightly elevated, longitudinal ridges
of sand, made up of gentle risings, disposed in a
crescent-like manner ; these have been produced
by the rills which flow back into the sea, or river,
at low water. In other examples, the surface is
marked by angular ridges irregularly crossing each
other, like the fissures in septaria ; these have obviously been caused by deposition into crevices produced in sand or mud by desiccation. A considerable
portion of stone, the flat, as well as the furrowed
surfaces, is covered with small, subcylindrical markings, which are the trails formed by some species
of vermes, or mollusca. The deepest furrows have
generally a slight coating of bluish clay, charged
with minute portions of lignite, and other vegetable
matter ; an appearance which has been occasioned
by streams from the shore which have flowed over
and coated the rippled sand. The furrowed sandstone presents an interesting example of the perfect
similarity of a natural process in periods separated
from each other by immense intervals of time.*

* For a particular account of the Wealden strata in the
south-east of England, see Geology of the South-East of England. For their nature and distribution in Wilts, &c. see Dr.
Fitton's Memoir.

34. WEALDEN OF THE ISLE OF WIGHT.—Deposites partaking of the character of those I have described, appear at the back of the Isle of Wight, and form the lowermost strata throughout the southern half of the island. Clay, identical with the Weald clay, and containing Sussex marble, may be seen at Sandover bay, within a few hundred yards of the chalk, and extending into Red-cliff; and also at the junction, on the east of Fresh-water bay, where the clay abounds in the minute *shields* of *Cyprides*. At Brook-point, the cliffs, which are about thirty feet high, are formed of clay, with inferior beds of soft sandstone; and contain lignite and vegetable remains, strongly impregnated with pyrites. Trunks of trees, of a coaly blackness, are seen imbedded in the clay of the cliff, and scattered on the shore. In many of the stems the ligneous structure is beautifully preserved, and veined with pyrites—other portions resemble jet. The strand at low water is seen to be formed of these fossils; and upon removing the sea-weeds which grow on the shore, the petrified trees occur imbedded in masses of clay, which have become indurated, and are now in the state of an argillaceous rock: the stems are from one to two feet in diameter, and eight or ten feet long. The knotty bark and ligneous fibre are very distinct.* Enormous bones are frequently found along the shore at

* From Mr. Webster's interesting account of the Geology of the Isle of Wight.

Sandover-bay and Brook-point, being washed out
of the beds of the Wealden, which there form part
of the basin of the British Channel.

35. ISLE OF PURBECK.—The Isle of Purbeck,
on the Dorsetshire coast, is of an irregular oval
form, about twelve miles in length, and seven in
breadth. On the eastern promontory, the chalk is
vertical, and beds of clay, sandstone, and limestone
are seen underlying the displaced strata of the
chalk formation; towards the southern extremity of
the island, the Portland limestone appears.*

Purbeck has long been celebrated for its quarries,
which have been worked from time immemorial,
and particularly during the middle ages; the compact varieties of limestone, which bear a good
polish, having, under the name of Purbeck marble,
been in great request for the religious edifices of
that period; and there is scarcely a cathedral, or
ancient church in the kingdom, that is not ornamented with columns, pavements, or sepulchral
monuments, constructed of this material. The
Purbeck limestone abounds in organic remains;
and the marble is a congeries of small fresh-water
snail-shells (*Paludina*), intermixed with the minute
crustaceous coverings of a species of *Cypris*. How
interesting is the reflection, that the beautiful cluster-
columns, the richest ornaments of Chichester cathedral, are entirely composed of the shelly coverings

* Conybeare and Phillips.

of snails which lived in the river of a country inhabited by colossal reptiles!

The vertical position into which so considerable a portion of the strata has been thrown, gives rise to interesting sections in the coves on the western side of the island ; and, in the precipitous cliffs of those basins, the Chalk, Weald, Purbeck, and Portland strata, although vertical, may be seen in their regular order of succession. No fewer than nine sections of the beds between the chalk and Portland stone (the upper division of the oolite, of which I shall speak in the next lecture) are visible on the shore, within the short space of five miles, in the small bays by which the coast is indented.*

36. ISLE OF PORTLAND—PETRIFIED FOREST.— The island, or peninsula, of Portland is a bold headland, off Weymouth, about four miles and a half in length, and two in breadth, united to the mainland by the Chesil beach. It presents a precipitous escarpment on the north, and, declining towards the south, appears, on approaching it from the Dorchester coast, like an inclined plane, rising abruptly from the ocean. The southern extremity is flanked by low calcareous cliffs, which, from the constant action of the sea, are worn into hollows and caverns. The base of the island is formed of a blue clay (*Kimmeridge clay*), surmounted by thick beds of the oolitic limestone, known as the Portland stone,

* Dr. Fitton, p. 215.   See Conybeare and Phillips' Geology of England, p. 159.

and which are extensively quarried on the northern brow of the island.

On this oolitic limestone are fresh-water strata
(the lowermost beds of the Wealden formation),
which are characterised by phenomena of the

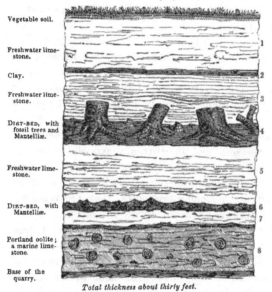

Vegetable soil.

Freshwater lime-
    stone.   1

Clay.   2

Freshwater lime-
    stone.   3

DIRT-BED, with
    fossil trees and
    Mantelliæ.   4

Freshwater lime-
    stone.   5

DIRT-BED, with
    Mantelliæ.   6
                                                  7

Portland oolite;
    a marine lime-
    stone.   8

Base of the
    quarry.

Total thickness about thirty feet.

TAB. 42.—SECTION OF A QUARRY IN THE ISLE OF PORTLAND.

(*From Dr. Fitton's Memoir on the Strata below the Chalk.*)

highest interest. Mr. Webster, in his admirable
geological memoir on the Isle of Wight, first directed attention to these remarkable deposites.
Upon the upper layer of marine limestone (8),

which abounds in ammonites, trigoniæ, and other characteristic shells of the oolite, is a fresh-water limestone, covered by a layer of *bituminous earth, or vegetable mould* (4), which (as you may perceive from these specimens, collected a few years since) is of a dark brown colour, contains a large proportion of earthy lignite, and, like the modern soil on the surface of the island, many water-worn stones. This layer is termed the *Dirt-bed*, by the workmen; and in and upon it are trunks and branches of coniferous trees, and plants allied to the recent *Cycas* and *Zamia*. Many of the trees, as well as the plants, are still erect, as if petrified while growing undisturbed in their native forests; having their roots in the soil, and their trunks extending into the upper limestone (see Tab. 42). As the Portland stone lies beneath these strata, which are not much used for economical purposes, they are removed, and thrown by as rubbish. On my visit to the island in the summer of 1832, the surface of a large area of the dirt-bed was cleared, preparatory to its removal, and a most striking phenomenon was presented to my view. The floor of the quarry was literally strewn with fossil wood: and I saw before me a petrified, tropical forest, the trees and the plants, like the inhabitants of the city in Arabian story, being converted into stone, yet still maintaining the places which they occupied when alive! Some of the trunks were surrounded by a conical mound of calcareous matter, which had evidently once been earth, and had accumulated around the bases and roots of the trees. The stems were generally three or four feet high, their summits being jagged and splintered, as if they had been torn and wrenched off by a hurricane,—an appearance which many trees in this neighbourhood, after the late storm, strikingly resembled. Some of the trunks were two feet in diameter ; and the united fragments of one tree measured upwards of thirty feet in length ; in other specimens, branches were attached to the stem. In the *dirt-bed*, there were many trunks lying prostrate, and fragments of branches. The fossil plants are called Cycadeoidea by Dr. Buckland, from their analogy to the recent Cycas and Zamia,* but for which M. Adolphe Brongniart has established a new genus, named *Mantellia*. The plants occurred in the intervals between the trees; and the dirt-bed was so little consolidated, that I dug up with a spade, as from a parterre, several specimens that must have been on the very spot in which they grew, having, like the columns of Puzzuoli, preserved their erect position (Tab.42.) amidst all the revolutions which the surface of the earth has

---

* These plants are so common in green-houses, that their forms must be well known. In the conservatories of the Coliseum, in the Regent's Park, are fine examples of the Dracæna, Yucca, Cycas, and several species of palms, allied to the fossil plants of Tilgate Forest. The magnificent collection of palms of the Messrs. Loddige, of Hackney, is referred to in my work on the Fossils of Tilgate Forest.

subsequently undergone, and beneath the accumulated spoils of countless ages. The trees and plants
are completely petrified by silex, or flint : you perceive that sparks are emitted upon striking a piece of
steel with this fragment, of what was once a delicate
plant. I may observe, that the common forms of
the fossil *Cycadeæ* (*Mantellia nidiformis* of Brongniart), are called *crows' nests* by the quarry-men.
I must not dwell longer on these extraordinary
phenomena, but refer you to the memoirs of Mr.
Webster, Dr. Fitton, and Dr. Buckland. From
what has been stated, it is evident, that after the
marine strata forming the base of the Isle of
Portland were deposited at the bottom of a deep
sea, and had become consolidated, the bed of that
ancient ocean was elevated above the level of the
waters, became dry land, and covered with forests.
How long this new country existed, cannot be ascertained; but that it flourished for a considerable
period is certain, from the number and magnitude
of the trees of the petrified forest. In the Isle of
Purbeck, traces of the dirt-bed, with the trunks of
trees, are seen beneath the fresh-water limestones
of the Weald ; a proof, that before the deposition
of the Purbeck marble could have taken place, the
petrified forest must have sunk to the depth of
many hundred feet.

37. FOSSILS OF THE WEALDEN.—The organic
remains of the Wealden consist of leaves, stems
and branches of plants of a tropical character,

bones of enormous reptiles of extinct genera, of crocodiles, turtles, flying reptiles, and birds; fishes of several genera and species, and shells of a fluviatile character. The bones are, for the most part, broken and rolled, as if they had been transported from a distance. They are strongly impregnated with iron, and are commonly of a dark-brown colour; their cavities are frequently filled with white crystallized carbonate of lime. The specimens in the loose sand and sandstone are often porous and friable; those in the Tilgate grit, heavy, brittle, and well preserved. In fractured portions imbedded in the limestone, the interstices are filled with calcareous spar, and the cancellated structure of the bones is often permeated by the same substance. The fossil vegetables occur bituminized, and in the state of casts of sandstone; the stems and branches are sometimes silicified; carbonized leaves and twigs are abundant in some of the strata. The shells in the clays have undergone but little change, and in many examples, the epidermis still remains; in the limestone, the substance of the shell is converted into spathose carbonate of lime. With these general remarks, I pass on to the enumeration of the principal organic remains.

38. Fossil Vegetables.—From the abundance of carbonaceous remains of vegetables in many of the laminated shales and clays of the Wealden, and the occurrence of lignite, or brown-coal, in masses and layers, which sometimes alternate

z

with beds of stone abounding in fresh-water bivalves, a striking analogy is presented to some of
the divisions of the *coal measures ;* and many years
since, this resemblance gave rise to a search for
coal at Bexhill, which, of course, proved abortive.*
But notwithstanding the prevalence of vegetable
matter in the strata, specimens exhibiting the nature
of the original plants, in any tolerable degree of
preservation, are rare; and, although my researches
have been unremitting, I have obtained but few
fossil plants that will admit of satisfactory conclusions as to their original structure.   I shall defer to
the lecture on fossil botany, a particular description
of these remains, confining my present remarks to a
brief account of the principal varieties, and the circumstances under which they occur.

   Entire layers of the calciferous grit of Tilgate
Forest are so full of minute portions of carbonaceous matter,† as to acquire a dark mottled colour ;
upon examining the imbedded particles, they appear
to be the detritus of plants ground to pieces by
agitation in sand and water.   Specimens in my possession, show that the greater part have been derived
from two elegant ferns, of extinct species, which are
peculiar to the Wealden.   The one is characterised
by its slender and minutely divided leaflets (*Sphenopteris Mantellii*—Tilgate Fossils, Plate 3) ; the

---

* Geology of the South-East of England, p. xviii.   Fossils
of the South-Downs, p. 35.
   † Fossils of Tilgate Forest, Plate 3, Fig. 6.

other by the distribution of the nervures of the
leaves (*Lonchopteris Mantellii* — Tilgate Fossils,
Plate 3), as you may perceive in the specimens
before you. This plant has been found in the
valley of Bray by M. Graves of Beauvais, and

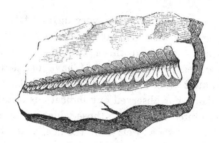

Tab. 43.—Leaflet of Lonchopteris Mantellii: Tilgate Forest.

(A Fossil Fern, peculiar to the Wealden.)

in the lower Boulonnais in France, and in Sweden, in strata of the same epoch as the Wealden.
These ferns probably did not exceed a few feet in
height; I have one stem of the *Sphenopteris* which
indicates a plant of about five or six feet. Several
other species are associated with these remains; but
the two plants I have named constitute by far the
greatest proportion of the fossil vegetables of Tilgate
Forest. Leaves of *Cycadeæ*, and seed-vessels of a
species of *Restiacea*, occur in the ironstone of
Heathfield; they are supposed to be of the same
species as specimens from Bornholm. Among the
z 2

plants from Heathfield are impressions which bear a close resemblance to those of the foliage of the cypress; while others appear to be referrible to fuci. The stalks of a species of mare's-tail (*Equisetum Lyellii**) abound in the blue limestone of Pounceford.

The stems of two plants, very distinct from each other, are the only vegetables of any considerable magnitude that occur throughout the Wealden of the south-east of England. I have not detected the slightest trace of wood like that of the forest of Portland, nor observed any indications of drifted and perforated masses like those which are so common in the sands and clays of the chalk, and other formations.

In my first publication on the fossils of Tilgate Forest, I described the plants which I now place before you. The first species consists of stems, with numerous tubular cavities, lined with quartz crystals, and presenting a structure decidedly analogous to the Cacti, or Euphorbiæ ; they have an external coating of carbonaceous matter, and, on the removal of this coaly crust, the outer surface has a remarkably eroded appearance. The stems vary from a few inches to two feet in circumference ; I have seen fragments which, when united, gave a length of five feet. There are no indications of branches, but many of the specimens taper at each end, and are of a clavated form, as in some species of cactus.

* Geology of the South-East of England, p. 245.

Dr. Fitton describes an interesting assemblage of these stems, which he fortunately observed before their removal, imbedded in clay, in a cliff to the east of the white-rock at Hastings : they were lying with their largest diameter in a horizontal position, and consisted of a silicious stem or nucleus, coated by lignite, which not only invested the stem, but also extended beyond each extremity. The stems, when cut and polished, exhibit the monocotyledonous* structure ; Count Sternberg considers

TAB. 44.—PORTION OF THE STEM OF CLATHRARIA LYELLII.
TILGATE FOREST.

them related to the palms. This fossil vegetable, from the characters above specified, has been named Endo-genites erosa. The other plant (*Clathraria Lyellii*†

* Dr. Fitton has given beautiful engravings of these fossils in the Memoir already cited, Plates XIX. and XX.

† The structure of this plant will be described more minutely in a succeeding lecture.

bears an analogy to the Yucca, and Dracæna or dragon-blood plant. Stems, with the markings of the bases of the leaves, point out the relation of this vegetable to the arborescent ferns,* while its internal structure is essentially different. This interesting specimen exhibits an internal axis, surrounded by a false bark, the surface of which is scored with the markings derived from the attachment of the leaf-stalks. The Clathraria has only been found in the quarries in Tilgate Forest. I have fragments of stems indicating a large size.

SEED-VESSELS.—Not only are the stems and leaves of plants and trees preserved in the Wealden beds, but even very delicate seed-vessels are sometimes found in the grit and sandstone. A small oval carpolithe (Tab. 45, fig. 1) is the most common. M. Adolphe Brongniart considers it probable, that this may belong to the *Clathraria Lyellii.*† The seed-vessels of coniferous trees also occur. These drawings (figs. 2, 3, 4) are from specimens belonging to my friend, Dr. Fitton, who has figured and described them in the valuable Memoir to which I have so often referred. Figs. 2 and 3 are copied inaccurately ; but I particularly claim your attention to fig. 4, which is half the size of the original ; this

---

* The reader may form an idea of the height and proportions of these elegant trees, by inspecting a specimen of tree-fern, forty-five feet high, from Bengal (*Alsophila Brunoniana*), on the staircase of the British Museum.

† See Geology of the South-East of England, p. 246.

beautiful cone was found imbedded in grit, in a quarry on Ashdown Forest, on the estate of Henry Shirley, Esq., and is remarkable for the double prominences on the scales.

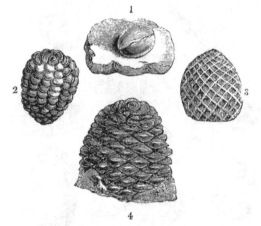

TAB. 45.—FRUITS AND CONES OF THE WEALDEN.

Fig. 1. *Seed-vessel of Clathraria Lyellii.* 2. *Cone from the Isle of Purbeck.* 3. *Cone from Kent.* 4. *Cone from Pippingford.**

39. FOSSIL SHELLS.—The shells of the Wealden, a series of the principal species of which I have placed before you, belong to but few genera; and although whole tracts of country are composed of their remains, and many of the limestones are mere

* Figs. 2, 3, 4, are reduced one-half from Dr. Fitton's Memoir, Pl. XXII.

conglomerates of shells, yet the species are but few; a character perfectly agreeing with that which prevails among the existing genera of our rivers. The

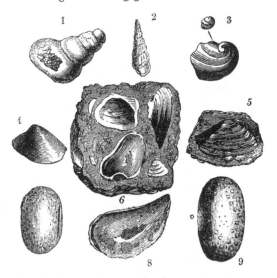

TAB. 46.—SHELLS AND CRUSTACEA OF THE WEALDEN.

Fig. 1. Paludina Sussexiensis, from a slab of Sussex marble. 2. Melanopsis attenuata. 3. Neritina Fittoni. 4. Cyclas. 5. Psammobia tellinoides. 6. Unio Walteri, with other shells. 7, 9. Cypris granulosus, highly magnified. 8. Mytilus Lyellii.

bivalve shells chiefly consist of muscles (referrible to a genus called *Unio* by conchologists), the casts of which are abundant in some of the sandstones; and several species of Cyclas,* that occur in myriads

* A genus of lacustrine, or freshwater bivalves.

in the shales and clays, and resemble tertiary shells in their state of preservation. The shales of Pounceford are very like the clays of Castle Hill, Newhaven, in respect to the layers of shelly remains which occur between the strata. The grey limestones are wholly composed of *Cyclades*, imbedded in argillaceo-calcareous cement, in which univalves are of rare occurrence.

The calciferous grit near Hastings is full of *Cyclades;* but the shells are decomposed, and the casts or impressions of the interior alone remain. In the argillaceous septaria of the Weald clay, casts of small univalves, also destitute of their shells, abound. In Langton-green quarry, near Tunbridgewells, layers of argillaceous rock inclose impressions of numerous shells; and among others a remarkable species of Unio, (Tab. 46, fig. 6.) This spiral univalve, *Melanopsis attenuata* (Tab. 46, fig. 2), belongs to a fresh-water genus; it occurs in the shale at Pounceford, in a beautiful state of preservation. This minute and elegant shell (Tab. 46, fig. 3,) is dispersed among small snail-shells in the grit of Tilgate Forest; I have named it *Neritina Fittoni*, in honour of my friend Dr. Fitton, whose able investigations have so fully elucidated the geological character and relations of the deposites below the chalk.

40. SUSSEX MARBLE; FOSSIL CYPRIDES.—The Weald-clay throughout its whole extent contains beds of limestone made up of a few species of

the univalve, called *Paludina* (Tab. 46, fig. 1), a
fresh-water snail, common in rivers and lakes. The
shells are sometimes decomposed, and their casts
alone remain, the interstices being filled up with indurated marl, or calcareous concretions. In the
coarser varieties are cavities left by the decomposition of the shells; in the compact masses the
whole has been permeated with a crystalline calcareous infiltration, of various shades of grey, blue,
and ochre, interspersed with pure white; polished
slabs, displaying innumerable sections of the inclosed shells, rival in interest and beauty many
foreign marbles. In these specimens you perceive
the shells in relief on one side, and sections of the
inclosed remains on the opposite polished surface ;*
very few bivalves occur in this limestone, which,
from its abundance in Sussex, is commonly known
by the name of *Sussex marble*. The Petworth marble,
and Bethersden stone of Kent, are from extensions
of the same beds. In western Sussex, blocks of a
beautiful marble mottled with green, blue, and
grey, have been found; it is composed of large
bivalves (*Unio*), and interspersed with a few univalves and fragments of reptiles, bones, &c. The
Purbeck marble, already described, only differs from
that of Sussex in the size of the shells; the paludinæ
in that limestone being of a very small species.

CYPRIS.—I have stated that the Wealden marbles

* Geology of the South-East of England, p. 184; ibid. p. 254.

are principally composed of fresh-water shells; but
other animal remains enter into their composition,
and which although so minute as to elude common
observation, possess a high degree of interest. It has
been mentioned that certain crustaceous animals,
(*Cypris*) abundant in fresh-water, having their
bodies protected by shells or cases which they shed
annually, occur in a fossil state in the lacustrine
tertiary deposites (page 231); and I referred to the
exhibition of the oxy-hydrogen microscope in illustration of the forms of the living species. The
shields of various kinds of these microscopic creatures abound in the Wealden clay,* septaria, and
limestones; and entirely fill up the interstices between the shells of some varieties of Sussex marble.
In these shales from near Lewes, septaria from
Barcombe, and marble from Laughton, by the aid
of a lens, hundreds of the shells of the cypris may
be detected. Dr. Fitton, who has investigated the
nature of these minute relics with his accustomed
acumen, has discriminated several species. These
enlarged drawings, from his illustrations, represent
a variety in which the shells are studded with
tubercles (Tab. 46, figs. 7, 9.) The natural size of
these objects does not, as you observe in the
specimens, exceed that of a pin's head, yet in certain formations entire layers of stone are composed
of their consolidated remains, and they constitute

* Dr. Fitton's Memoir, Pl. XXI. figs. 1, 2, 3, 4.

a large proportion of the mass of many beds of Sussex marble.

41. Fishes.—Bones, teeth, rays, and detached scales of fishes of the shark family, and of species allied to the large river-pikes of South America, are very abundant; but rarely any united portions of the skeleton, or scaly covering, are preserved; a circumstance arising from the drifted character of the Wealden deposites. Strong, thick, enamelled, lozenge-shaped scales, possessing a high polish, and having two processes of attachment, are very abundant in the sandstones, grit, and clays throughout the Wealden. At St. Leonard's and Tilgate Forest, the conglomerate contains immense numbers, associated with small hemispherical teeth, called *fishes'-eyes* by the workmen. These scales belong to two species of *Lepidotus*, or bony-pike, of which genus a recent species inhabits the rivers of South America. It is rarely that any considerable number of the scales remain attached to each other in their natural position; but I have a few specimens in which large portions of the scaly covering retain their original character. These fishes must have attained a large size. In a specimen presented to me by my liberal friend, Robert Trotter, Esq. F. G. S. of Borde Hill, near Cuckfield, a large mass of the united scales is beautifully preserved; it belongs to that part of the body where the caudal fin commences, is twelve inches wide, and must therefore have belonged to a fish

ten or twelve feet long, and three feet wide. Tri-
cuspid teeth finely striated, and fin-bones of five or
six species of genera belonging to the shark family,
are of frequent occurrence. The fishes of the
Wealden are entirely distinct from those of the
Chalk.*

42. REPTILES OF TILGATE FOREST.†—It will
doubtless excite your surprise to learn that the
whole of the enormous bones, and teeth, I have
placed on the table, belong to reptiles ; and that not
a vestige of the mammalia occurs in the Wealden.
Even these teeth, which so strikingly resemble the
incisors of the rhinoceros, and these bones of the
feet and toes, so similar in their construction and
magnitude to those of the hippopotamus, all belong
to oviparous quadrupeds ! Many of the specimens
before you can be referred to certain extinct forms
of Saurians ; but others are yet undetermined, in consequence of my want of leisure, and distance from any
extensive collection of subjects of comparative anatomy. The study of the fossil bones of the Wealden

* The following fishes of the Wealden, in my museum, have
been named by M. Agassiz. Pycnodus microdon, Lepidotus
Fittoni, L. Mantellii, Hybodus grossiconus, H. marginalis,
H. polyprion. A small species, *Lepidotus minor,* occurs in the
Purbeck limestone.

† This lecture was illustrated by several hundred specimens
of bones and teeth of reptiles from the Wealden ; many of such
enormous size, that the assemblage resembled an accumulation
of the disjointed skeletons of gigantic elephants or mastodons.
—G. F. R.

is indeed no easy task; for while in many marine
deposites, considerable portions of the skeletons, or
even the entire forms, are often discovered; in the
Wealden, with the exception of but three or four
examples, every bone, tooth, and scale, has been
found apart from each other; and as if to render the
task still more perplexing, the relics of several different species are scattered, as it were, at random
through the rocks. Every specimen, as I have
before remarked, bears evidence of having been
transported from a distance; it would seem as if
the limbs and carcases of the animals were floated
down the stream, and rolled backwards and forwards
by the tides, and the bones broken before they became imbedded in the mud of the delta. " To collect
these scattered fragments, and extricate them from the
solid rock; to reunite them into a whole, and assign
to each skeleton of the respective animals, the bones
which once belonged to it, yet not to confound the
different species together—such is the labour which
the comparative anatomist has to perform, who
undertakes to investigate the structure of the
Wealden reptiles." I reserve for the next lecture
some observations on the economy and habits of
the reptile tribes, and will now describe the fossil
relics before us.

43. Fossil Turtles.—The bones and plates of
turtles are very common in the Purbeck limestone,
and in the grit, sandstone, and shale of Tilgate
Forest. They are referrible to two or more fresh-

water, and one marine species; the former appear to be analogous to an *Emys*, or fresh-water turtle, described by Cuvier,* that occurs in the Jura limestone at Soleure. It is a very flat species, and probably attained two feet in length. The ribs of a *Trionyx* (which is also a fresh-water turtle), occur in the shale of Pounceford, and grit of Tilgate; the surface of the ribs or dorsal plates is shagreened all over, as is usual in these turtles, which have no shelly covering, but only a thick, tough skin, or integument.†

44. PLESIOSAURUS. Several bones, and vertebræ of the neck and back of the extraordinary extinct reptile, called *Plesiosaurus*,‡ whose remains are found in such prodigious numbers in the lias, occur in the calciferous sandstone of Tilgate Forest, and prove that this animal was an inhabitant of the sea into which the river of the Wealden flowed.

45. FOSSIL CROCODILES.—The skeletons of the crocodiles, alligators, and gavials, those well-known reptiles of Egypt, India, and America, possess characters which render their fossil bones easily recognisable by the comparative anatomist. The peculiar structure of the teeth, as you may observe in this specimen, (Tab. 47,) affords certain indications of the original animal. The teeth of the crocodile consist of a succession of cones, like a series of

---

* Oss. Foss. Tom. V. p. 232.

† Geology of the South-East of England, p. 255.

‡ Plesiosaurus—*akin to a lizard.*

thimbles of various sizes, fitted into each other; they
are striated externally, and have a prominent lateral ridge; as the outer tooth wears away, a

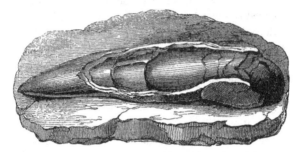

TAB. 47.—TOOTH OF CROCODILE—TILGATE FOREST.

new one is ready to supply its place; hence the
teeth of the old crocodile are as fresh as those of
the young animal but just escaped from its shell.
The example before us, from the removal of a portion of the outer surface, offers an instructive
example of the internal structure.   Detached bones
of several species of crocodiles are scattered through
the Tilgate strata.   From the difference observable
in the form of the teeth they appear referrible to
two kinds—the one belonging to that division of
crocodiles, with long slender muzzles, named *Gavials ;* the other to a species of Crocodile, properly
so called,* and resembling a fossil species found at

* Geology of the South-East of England, p. 263.

Caen. Among the hundreds of teeth and bones, collected in the Wealden, no portions of the jaw, or any united parts of the skeleton, have been observed. In the Purbeck beds a specimen has recently been discovered that affords an interesting illustration of the osteology of one of the Tilgate species.

46. THE SWANAGE CROCODILE.—Swanage, or Swanwich, is a little town on the east of the Isle of Purbeck, the inhabitants of which carry on a brisk trade in the exportation of stone from the numerous quarries in its vicinity, there being a fine bay and good anchorage for vessels. The town stands at the mouth of the bay, about six miles E. S. E. of Corfe Castle. The section, exposed by the coast, explains the geological structure of the country, and presents the following series :—First, beds above the chalk; secondly, chalk; and lastly, the fresh-water strata beneath; the Purbeck limestone occupying the lowermost place in the series. I need not, in this place, dwell on the dislocations of these strata, and the causes by which they have been disconnected and thrown into their present position. A few months since the workmen employed in a quarry, in the immediate vicinity of Swanage, had occasion to split asunder a large slab of limestone, when, to their great astonishment, they perceived many bones and teeth of some animal, on the surfaces they had just exposed. As this was no ordinary occurrence, for although scales of fishes, shells, &c.

were frequently observed in the stone, bones had never before been noticed, both slabs were carefully preserved by the proprietor of the quarry; and fortunately my intelligent friend, Robert Trotter, Esq. F.G.S., of Borde Hill, Sussex, happening to visit Swanage a short time afterwards, heard of the discovery, and with that liberality and ardour for the advancement of science for which he is distinguished, purchased the specimens, and presented them to my collection. I have cleared away the stone, so far as the brittle state of the bones will permit without injury, and they are now rendered two as interesting groups of Saurian remains as exist in this country.

In these specimens a considerable portion of the left side of the lower jaw, with two teeth attached, is preserved; many teeth are scattered over the stone, and numerous *dermal*, or skin-bones, which are readily distinguished, not only by their form, but also by their deeply pitted surface. The pelvis is nearly entire, and there are many bones of the spine, (*caudal, and dorsal vertebræ*, and *chevron-bones*,) ribs, and some of the long bones of the extremities.

47. THE MEGALOSAURUS.*—The fissile limestone of Stonesfield, of which I shall speak in the next lecture, has long been celebrated for the teeth and bones of a gigantic reptile, to which Dr. Buckland

---

* Megalosaurus,—*great lizard.*

has given the name of *Megalosaurus.* In this place
I will only state, that vertebræ, bones, and teeth of
this animal have been found in the Tilgate grit,
and in the clays and sandstones of the Wealden,
associated with the remains of turtles, crocodiles,
and of the still more colossal oviparous quadruped
the *Iguanodon,* which I now proceed to notice.

48. THE IGUANODON.—It is now several years,
since the discovery of a mutilated fragment of a
tooth, led me to suspect the existence of a gigantic
herbivorous animal in the strata of Tilgate Forest,
which subsequent researches confirmed.* This is the
fragment: it is part of the crown of a tooth, resembling in its prismatic form the incisor of one of the
herbivorous mammalia, worn by use. The enamel
is thick in front and thin behind, and by this disposition a sharp cutting edge is maintained in every
stage. Here, then, is a character, which if we bear
in mind the principles of comparative anatomy enforced in the second lecture, (page 128,) will afford
us certain indications as to the nature of the animal
to which it belonged. The structure of the tooth,
and its worn surface, prove that it is referrible to a
species that fed on vegetables ; the absence of a fang,
and the appearance of the base, not broken, but
indented, show that the shank has been absorbed
from the pressure of a new tooth, which has grown

---

* " On the teeth of the Iguanodon, a newly-discovered fossil
herbivorous reptile, from the strata of Tilgate Forest."—*Philos.
Trans.*

up and supplanted the old one; a process too
familiar to require explanation.*

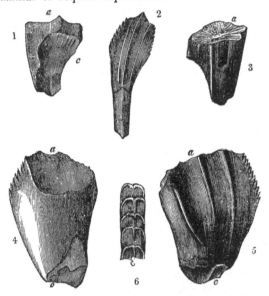

TAB. 48.—TEETH OF THE IGUANODON, FROM TILGATE FOREST.

Figs. 1, 3. *Tooth worn flat, and the fang absorbed.* 2. *Tooth of a young
animal.* 4. *Inner, and* 5, *outer, surface of a tooth of an adult.* 6. *Lateral view of the serrated edge of fig.* 5, *magnified.* a, *The surface worn by
mastication.* c. *The indentation produced by pressure of the new tooth.*

* It cannot be requisite to notice the vulgar error that the
first teeth in children have no fangs; it may however elucidate
the remarks in the text, if the reader is reminded that the absence of fangs in the teeth shed in childhood, results from the
absorption of the fang of the old teeth, occasioned by the pressure of those which are to supply their place.

In the teeth before us we trace every gradation of this change, from the perfect form (Tab. 48, fig. 2,) — the partially worn specimen, (figs. 4, 5,) to the mere stump, (figs. 1, 3,) in which the crown is worn flat, and the absorption of the fang complete. The teeth, when perfect, are of a prismatic form, and remarkable for the prominent ridges which extend down the front, and the serrated margins of the crown. (Tab. 48, fig. 6, the serrated edge magnified.)

But although by this mode of induction the grand division of the animal kingdom to which the original belonged was determined, a rigid comparison of the teeth with those of recent species was necessary, to arrive at more satisfactory results. In a fossil state, no teeth at all analogous had been noticed: and after a fruitless research through the collections of comparative anatomy in London, I found, in the jaws of a recent Iguana,* the type for which I had so long sought in vain. The Iguanas are land lizards, natives of many parts of America and the West Indies, and are rarely met with north or south of the tropics. They are from three to five feet in length, and feed on insects and vegetables, climbing trees, and chipping off the tender shoots. They nestle in the hollows of rocks, and deposit their eggs, which are like those of

* Prepared by Mr. Stutchbury, the intelligent curator of the Bristol Institution.

turtles, in the sands or banks of rivers. The
Iguana is furnished with a row of very small,
closely-set, pointed teeth, with serrated edges, which
are attached at the base, and by the outer surfaces of the fangs, to the jaw, the alveolar process
forming an external parapet; there is no internal
bony covering. The new teeth arise at the base
of the old, and supplant them by occasioning
the absorption of the fangs. The teeth of the
Iguana closely resemble the perfect tooth, Tab. 48,
fig. 2, except in size; those of the recent animal
scarcely exceeding in magnitude the teeth of the
common mouse. But in the Iguana the teeth
never present a worn surface; they are broken or
chipped off by use, but not ground smooth as in
the herbivora. The reason is obvious; none of the
existing reptiles are furnished with cheeks or moveable coverings to their jaws, and therefore cannot
perform mastication; their food or prey is seized
by the teeth and tongue, and swallowed whole.
But apart from this discrepancy, the teeth and
mode of dentition of the fossil animal are so perfectly analogous to those of the Iguana, that I
have named the original the IGUANODON, signifying
an animal having teeth like the Iguana. In the
course of the present summer a portion of jaw has
been discovered, which confirms all the inferences
that many years since I ventured to deduce from
the teeth alone.

From the gigantic size of the fossil teeth, as com-

pared with the recent, I was led to infer that many
of the colossal bones, collected from time to time,
in Tilgate Forest, belonged to the same kind of
animal. By comparing the bones with the skeleton
of the Iguana, (presented me by Baron Cuvier,)
I succeeded in determining many parts of the skeleton ; and at length was enabled to restore, as it
were, the form of the Iguanodon, and ascertain its
proportions ; the correctness of my inferences was
shortly to be put to the test, by a discovery in a
neighbouring county.

49. The Maidstone Iguanodon.* — In May,
1834, some workmen employed in a stone-quarry,
in the occupation of Mr. W. H. Bensted, of Maidstone, observed in a mass of rock which they had
blasted, several portions of what they supposed to
be petrified wood ; they preserved the largest piece
for the inspection of Mr. Bensted, who at once perceived that it was a portion of bone belonging to
some gigantic animal. He therefore gave directions
that every fragment should be collected, and after
much labour and research, succeeded in obtaining
those pieces, which are now united, and form a
specimen of the highest interest; he also cleared
away part of the surrounding stone, so as to expose
the bones, which I have since completely developed
and joined together.

The specimen consists of a considerable number
of the bones, composing the inferior portion of the

* Appendix.

skeleton of an Iguanodon, which, when living, must have been upwards of 60 feet in length. The bones are imbedded in the stone in a very confused manner, few of them being in their natural order of juxta-position, and all more or less flattened and distorted. The following are well displayed ; and there are many fragments of others, which are too imperfect to admit of being determined.

Two *thigh-bones*, each 33 inches long.[*]

One *leg-bone* (*tibia*), 30 inches long.

Metatarsal and phalangeal bones of the hind feet; these much resemble the corresponding bones in the Hippopotamus.

Two *claw-bones* (*unguical phalanges*), which were covered by the nail or claw; these correspond with the unguical bones of the land tortoise.

Two finger, or metacarpal bones of the fore-feet, each 14 inches in length.

A *radius*, or bone of the fore-arm.

Several *dorsal* and *caudal vertebræ* (bones of the spine and tail).

Fragments of several ribs.

Two *clavicles*, or collar-bones, each 28 inches in length, resembling the bone figured Plate IV. figs. 1, 2, Geology of the South-East of England. These bones are of a very singular form, and differ essentially from any known clavicle, yet it seems impossible to assign them to any other place in the skeleton.

Two large flat hatchet-shaped bones, which appear to belong to the pelvis, and are probably the *ossa ilia*.

---

[*] The femur of the Iguanodon is very remarkable; it has a large trochanter opposite to the head of the bone, and a process on the inner side for the attachment of powerful adductor muscles; the front of the lower extremity is deeply grooved anteriorly, as in the toad; the shaft of the bone is sub-quadrangular.

A *chevron-bone*, or one of the inferior spinous processes of a vertebra of the tail.

A portion of a tooth, and the impression of another.—The preservation of these teeth is most fortunate, as the identity of the animal with the *Iguanodon* of Tilgate Forest is thereby completely established.

The stone in which the bones are imbedded is of that hard variety of the grey, arenaceous limestone, called *Kentish rag*, which is much employed in various parts of Kent, and in the west of Sussex, for building, and repairing roads. This *Rag* belongs to the Shanklin sands, and abounds in the marine shells which are characteristic of that division of the chalk formation. In the quarry in which the remains of the Iguanodon were found, Mr. Bensted has discovered fossil wood perforated by *lithodomi*, or boring shells; impressions of leaves, stems of trees, *Ammonites, Nautili*, &c.; large conical striated teeth, which are referrible to those extinct fossil fishes, which M. Agassiz denominates *Sauroid*, or lizard-like; scales and teeth of several kinds of fishes, and, among these, a jaw or mandible of that singular genus of fish, the *Chimera*.

The geological position of this specimen forms an exception to what has been previously remarked of the fossils of the Wealden; for while the bones in the latter are found associated with terrestrial and fluviatile remains only, the Maidstone specimen is imbedded in a marine deposit. This discrepancy, however, in no wise affects the arguments previously

advanced, as to the fluviatile origin of the strata of
the Wealden ; it merely shows that part of the delta
had subsided, and was covered by the chalk ocean,
whilst the country of the Iguanodon was still in existence. The body of an Iguanodon was then drifted
out to sea, and became imbedded in the sand of the
ocean ; in like manner, as at the present day, bones
of land quadrupeds may not only be engulfed in
deltas, but also in the deposites of the adjacent sea.

This specimen possesses a high interest, because
it proves that the separate bones found in the strata
of Tilgate Forest, and which I had assigned to the
Iguanodon, solely from analogy, have been correctly
appropriated ; and we obtain also a knowledge of
many interesting facts relating to the structure and
economy of the original. I can notice but one of
these inductions. As the Iguana lives chiefly upon
vegetables, it is furnished with long slender feet, by
which it is enabled to climb trees with facility, in
search of food. But no tree could have borne the
weight of the colossal Iguanodon, — its movements
must have been confined to the land and water, and
it is evident that its. enormous bulk must have required limbs of great strength. Accordingly we find,
that the hind feet, as in the Hippopotamus, Rhinoceros, and other large mammalia, were composed
of strong, short, massy bones, furnished with
claws, not hooked as in the Iguana, but compressed
as in the land tortoises ; thus forming a powerful
support for the enormous leg and thigh. But the

bones of the hands, or fore feet, are analogous to those of the Iguana, — long, slender, flexible, and armed with curved claws the exact counterpart of the nail-bones of the recent animal; thus furnishing prehensile instruments fitted to seize the palms, arborescent ferns, and dragon-blood plants, which probably constituted the food of the original. Here we have another interesting example of that admirable adaptation of structure to the necessities and conditions of every form of existence, which is alike manifest, whether our investigations be directed to the beings around us, or to the structure of those which have long since passed away.

Gigantic as must have been the animal discovered by Mr. Bensted, there are in my collection many bones which indicate yet more colossal proportions. A thigh bone, from the west of Sussex, (presented to me by J. Napper, Esq.) is 3 feet 8 inches long, and 35 inches round, at the largest extremity; and the shaft of another femur is 24 inches in circumference! The following is the result of a careful comparison of some of the fossil, with the corresponding bones of the Iguana, with the view of ascertaining the probable *average* size of the original animal, (*vide* Geol. of South-East of England, p. 315;) we should, however, bear in mind, that some individuals must have far exceeded this estimate, and, if they bore the proportions of the recent Iguana, have been upwards of 100 feet in length!

Length of the *Iguanodon*, from the snout to the

tip of the tail, 70 feet. Circumference of the body, 14½ feet. Length of the tail, 52½ feet. Length of the hind foot, 6½ feet. Of course this calculation is offered but as an approximation; we cannot, however, for a moment doubt, that an animal possessing such a body, as to require a thigh bone eight inches in diameter, must have been of prodigious magnitude; such a thigh-bone, if covered with muscles and integuments, would be upwards of seven feet in circumference!

I will notice one other remarkable feature in the structure of the Iguanodon. The Iguanas are distinguished among the lizards by their exuberant dermal appendages; some have serrated processes or spines on the back, (as in this specimen from Barbadoes, presented to me by my friend R. I. Murchison, Esq., late President of the Geological Society;) others on the tail; while many have warts and horny protuberances on the head and snout. The extraordinary relic I place before you, is the FOSSIL HORN of the Iguanodon, from Tilgate Forest.* It is composed of bone, and bears marks on its surface of the integument with which it was invested; it is four inches high; the base, which is of an irregular elliptical form, is 3.2 inches by 2.1. In this additional analogy between the Iguanodon and the Iguana, we perceive another instance of

---

* This fossil was discovered by Mrs. Mantell; see Geology of the South-East of England, page 312, Plate III. fig. 5. Dr. Buckland has copied this figure in his Bridgewater Essay.

that law of co-relation of form of which our researches have afforded so many examples.

I will only add, that the fossil plants with which the remains of the Iguanodon are associated, were furnished with tough, thick stems, like those of the palms, tree-ferns, yucca, &c. These probably constituted the food of the original; and the peculiar structure of its teeth was evidently required, and admirably adapted, for the mastication of such vegetable productions.

50. THE HYLÆOSAURUS (*Wealden Lizard*).— In the summer of 1832, I discovered, in Tilgate Forest, the remains of a reptile, not less extraordinary than the Iguanodon, and which I have named the *Hylæosaurus*, to denote its relation to the Wealden formation. A block of calciferous grit had been broken up by the quarry-men, and a great part of it thrown upon the road, as it was not supposed to contain any thing interesting. Accidentally visiting the quarry, I noticed indications of bones in several pieces of stone on the road-side, and therefore directed that the remaining portions should be collected, and sent to my residence. Having cemented the fragments together, and chiselled off the hard grit in which the bones were wholly imbedded, and to which they are still attached, I succeeded, after much labour, in displaying a considerable portion of the skeleton of a reptile, which *blends the osteology of the crocodile with that of the lizard.* The vertebræ of the neck, several of the

back, many ribs, and the bones of the *sternum*, or chest, remain; there are also *dermal*, or skin-bones, which, in animals of this family, support the large scales. But the most extraordinary parts, are many *enormous, angular, spinous bones*, which lie in the direction of the vertebral column, and evidently extended originally like a serrated fringe along the back of the animal. Many of the existing lizards have remarkable appendages of this kind, particularly the *Cyclura*. This figure, from Dr. Harlan's valuable work,* shows how largely these curious processes are developed in some species.

TAB. 49.—CYCLURA CARINATA.
(*A recent Lizard, allied to the Iguana. Dr. Harlan.*)

The length of the Hylæosaurus was probably about twenty-five feet. In the same block of stone were masses of vegetable remains, with seed-vessels, and stems of *Clathraria Lyellii.*† I have lately

* Medical and Physical Researches, by R. Harlan, M.D., F.G.S., 8vo. Philadelphia, 1835.

† Geology of the South-East of England, p. 316, Pl. V. is an excellent lithograph of the specimen, by Mr. Pollard, of West-street, Brighton.

obtained many bones of this extraordinary creature from a bed of clay near Crawley; and also a most interesting specimen of the vertebral column, comprising nearly thirty vertebræ of the tail and back, with many *dermal and spinous bones*, ribs, &c. exhibiting very peculiar osteological characters.

51. FLYING REPTILES, or PTERODACTYLES. — The remains of thin and slender bones, evidently adapted for an animal capable of flight, were among my earliest discoveries in the strata of Tilgate Forest. Some of these bones appear to belong to those singular extinct creatures, called *Pterodactyles*, or *wing-toed* reptiles, which had a beak like a bird, a long neck, and a wing, sustained principally on an elongated toe. It is sufficient merely to notice the occurrence of these remains in the Wealden; in the succeeding lecture the subject will be resumed.

52. FOSSIL BIRDS. — In describing the fossil remains of the animals of the older tertiary epoch, it was stated that several recent genera of birds were contemporaneous with the Palæotheria (page 220); but no traces of this class of animated nature had been found in the chalk, or in strata of an earlier date. The discovery of the undoubted remains of birds in the strata of Tilgate Forest, became, therefore, a fact of great interest and importance in the physical history of the globe. After selecting the bones which appeared to be referrible to Pterodactyles, several remained which bore so

striking a resemblance to those of *Waders*, that I ventured to describe them as such in my work on the Fossils of Tilgate Forest; and this opinion was corroborated by Baron Cuvier, to whom I showed the specimens on his last visit to England. Subsequently, I have obtained the inferior portion of a leg-bone (*tarso-metatarsal*), in which the oval cicatrix, or articulation for the hind-toe, is distinctly visible, and proves unquestionably that the bone belonged to some kind of wader, perhaps a heron; the position of the hind toe in birds, varying in accordance with the habits and economy of the respective orders * (page 123). These are the most ancient remains of this class of animals hitherto discovered.

53. THE COUNTRY OF THE IGUANODON.—By this survey of the strata and organic remains of the Wealden, we have acquired data from which, by the principles of induction explained in a former lecture (page 34), we may obtain secure conclusions as to the nature of the country from whence those spoils were derived, the animals by which it was inhabited, and the vegetables that covered its surface. That country must have been diversified by hill and dale, by streams and torrents, the tributaries of its mighty river. Arborescent ferns, palms, and

* See a Memoir "On the Bones of Birds discovered in the Strata of Tilgate Forest," by the author, Geological Transactions, 1838, which contains some interesting remarks of Professor Owen, on the osteological characters of the fossil bones.

yuccas, constituted its groves and forests, delicate ferns and grasses, the vegetable clothing of
its soil; and in its marshes, equiseta, and plants of
a like nature, prevailed. It was peopled by enormous reptiles, among which the colossal Iguanodon
and the Megalosaurus were the chief. Crocodiles
and turtles, flying reptiles and birds, frequented its
fens and rivers, and deposited their eggs on the
banks and shoals; and its waters teemed with
lizards, fishes, and mollusca. But there is no evidence that Man ever set his foot upon that wondrous
soil, or that any of the animals which are his contemporaries found there a habitation: on the contrary, not only is evidence of their existence
altogether wanting, but from numberless observations made in every part of the globe, there are
conclusive reasons to infer, that man and the existing
races of animals were not created, till myriads of
years after the destruction of the Iguanodon country
—a country, which language can but feebly portray,
but which the magic pencil of a Martin, by the aid
of geological research, has rescued from the oblivion
of countless ages, and placed before us in all the
hues of nature, with its appalling dragon-forms,
its forests of palms and tree-ferns, and all the luxuriant vegetation of a tropical clime.*

54. SEQUENCE OF GEOLOGICAL CHANGES.—Let
us review the sequence of those stupendous changes,

* See the Frontispiece; an engraving on steel, from an
original painting of John Martin, Esq., K L.

of which our examination of the geological phenomena of the south-east of England has afforded
such incontrovertible evidence. From the facts
brought before us, we learn that at a period incalculably remote, there existed in the northern
hemisphere an extensive island, or continent, possessing a climate of such a temperature, that its
surface was clothed with coniferous trees, arborescent ferns, and plants allied to the Cycas and Zamia; and that the ocean which washed its shores was
inhabited by turtles and reptiles of extinct genera.
This island and its forests suffered a partial subsidence, which was effected in such manner that many
of the trees, although torn and rent, still retained
their erect position; and the Zamiæ, and a considerable layer of the vegetable mould in which they
grew, remained undisturbed.   In this state an
inundation of fresh-water covered the once flourishing forest, and deposited upon the soil and around
the trees a calcareous mud, which gradually consolidated into fine limestone; water, holding flint in
solution, percolated through the mass, and silicified
the now submerged trees and plants.   A further
depression took place — a body of fresh water,
brought down by land-floods and rivers, overwhelmed the petrified forest, and heaped up accumulations of debris, which their parent streams
had washed away from the rocks over which they
had flowed. The country traversed by the rivers,
like that of the submerged forest, enjoyed a tropical

climate; it was clothed with palms, arborescent
ferns, and plants allied to the yucca and the dracæna, and tenanted by enormous reptiles, crocodiles, and land and fresh-water turtles; and in its
waters were various kinds of fishes, mollusca, and
aquatic plants. The bones and teeth of the reptiles
—the remains of the turtles—the teeth and scales
of fishes—the shells of the snails and muscles—
the stems, leaves, and even seed-vessels, of the
trees, were carried down by the stream, and deposited in the mud of the delta, beneath which the
petrified forest was now buried. This state continued for a long period: another change took
place; the country and its inhabitants were swept
away, and the delta, and the strata on which it
reposed, were submerged to a great depth, and
formed part of the bottom of a profound ocean,
whose waters teemed with myriads of zoophytes,
shells, and fishes, of species that are now no more.
Thermal waters, holding calcareous and silicious
matter in solution, were poured into its basin, and,
in its tranquil depths, layers of flint and chalk were
deposited. And so rapidly were these depositions
effected, that fishes, while in the act of swimming,
were arrested in their progress, and became suddenly
enveloped in a bed of rock. This epoch was of
considerable duration: at length elevatory movements began to take place, the bottom of the deep
was slowly up-heaved, and as the elevation continued, the depositions which had formed in the

basin of the ocean, and had become consolidated,
were broken up, and as they approached the surface
were acted upon by the waves; the chalk strata now
began to suffer degradation and destruction, till at
length the delta of the country of the Iguanodon
emerged above the waters, and finally, even the
ancient petrified forest was brought to view, and
became dry land. At length some masses rose to
an elevation of a few hundred feet above the level
of the sea, and formed a group of islands; but, in
the depressions of the strata beneath the waters,
deposites went on, from the waste of the cliffs on
the sea-shores. Large mammalia now inhabited such
portions of the former ocean-bed as were clothed
with vegetation, and as they died their bones and
teeth were enveloped in the sediments of mud and
gravel which were forming in the bays and estuaries. This era also passed away—the elevation
continued—other portions of the bed of the chalk-
ocean became dry land—and at length also those
newer deposites, in which the remains of the mammoth and the elk, the last tenants of the country,
were entombed.   The oak, elm, ash, and other
trees of modern Europe, sprang up where the
palms and tree-ferns once flourished; the deer,
boar, and horse, ranged where the mighty reptiles
had once ruled sole monarchs of the country; and
lastly, man appeared, and took possession of the
soil.   At the present time, a city stands on the
deposites which contain the remains of the elephant

and the elk; the huntsman courses, and the shepherd tends his flocks, on the elevated and rounded masses of the bottom of the ancient chalk-ocean; the farmer reaps his harvests upon the cultivated soil of the delta of the Iguanodon; and' the architect seeks, beneath the petrified forest, for the materials with which to construct his edifices.

55. RETROSPECT OF GEOLOGICAL ERAS. — Such is a plain enunciation of the results of our investigations; but I will embody these inductions in a more impressive form, by employing the metaphor of an Arabian writer, and imagining some higher intelligence from another sphere, to describe the physical mutations of which he may be supposed to have taken cognizance, from the period when the forests of Portland were flourishing, to the present time. Countless ages ere man was created, he might say, I visited these regions of the earth, and beheld a beautiful country of vast extent, diversified by hill and dale, with its rivulets, streams, and mighty rivers, flowing through fertile plains. Groves of palms and ferns, and forests of coniferous trees, clothed its surface; and I saw monsters of the reptile tribe, so huge that nothing among the existing races can compare with them, basking on the banks of its rivers and roaming through its forests; while, in its fens and marshes, were sporting thousands of crocodiles and turtles. Winged reptiles of strange forms shared with birds the dominion of the air, and the waters

teemed with fishes, shells, and crustacea. And
after the lapse of many ages I again visited the
earth; and the country, with its innumerable
dragon-forms, and its tropical forests, all had disappeared, and an ocean had usurped their place.
And its waters teemed with nautili, ammonites,
and other cephalopoda, of races now extinct; and
innumerable fishes and marine reptiles. And
countless centuries rolled by, and I returned, and,
lo! the ocean was gone, and dry land again appeared,
and it was covered with groves and forests; but
these were wholly different in character from those
of the vanished country of the Iguanodon. And I
beheld, quietly browsing, herds of deer of enormous size, and groups of elephants, mastodons,
and other herbivorous animals of colossal magnitude. And I saw in its rivers and marshes
the hippopotamus, tapir, and rhinoceros; and
I heard the roar of the lion and the tiger, and
the yell of the hyena and the bear. And another
epoch passed away, and I came again to the scene
of my former contemplations; and all the mighty
forms which I had left had disappeared, the face of
the country no longer presented the same aspect;
it was broken into islands, and the bottom of the
sea had become dry land, and what before was dry
land had sunk beneath the waves. Herds of deer
were still to be seen on the plains, with swine, and
horses, and oxen; and wolves in the woods and
forests. And I beheld human beings, clad in the

skins of animals, and armed with clubs and spears;
and they had formed themselves habitations in
caves, constructed huts for shelter, inclosed pastures
for cattle, and were endeavouring to cultivate the
soil. And a thousand years elapsed, and I revisited
the country, and a village had been built upon the
sea-shore, and its inhabitants supported themselves
by fishing; and they had erected a temple on the
neighbouring hill, and dedicated it to their patron
saint. And the adjacent country was studded with
towns and villages; and the downs were covered
with flocks, and the valleys with herds, and the
corn-fields and pastures were in a high state of
cultivation, denoting an industrious and peaceful
community. And lastly, after an interval of many
centuries, I arrived once more, and the village was
swept away, and its site covered by the waves; but
in the valley and on the hills above the cliffs a
beautiful city appeared; with its palaces, its temples,
and its thousand edifices, and its streets teeming
with a busy population in the highest state of civilization; the resort of the nobles of the land, the
residence of the monarch of a mighty empire.
And I perceived many of its intelligent inhabitants
gathering together the vestiges of the beings which
had lived and died, and whose very forms were
now obliterated from the face of the earth, and
endeavouring, by these natural memorials, to trace
the succession of those events of which I had been

the witness, and which had preceded the history of their race.*

* The concluding portion of these remarks refers to the changes that have taken place on the Sussex coast, during the historical era. Before the Conquest, the greater part of the little fishing town of Brighthelmston (*Brighthelm's-town*) was situated below the cliffs, on a terrace of beach and sand, now covered by the waves. The church, dedicated to St. Nicholas, the patron saint of fishermen, was placed on an eminence, that it might serve as a land-mark. The inroads of the sea led to the erection of buildings on the high ground, and its progressive encroachment gradually diminished the area of the ancient town, till at length a sudden inundation, but little more than a century ago, swept away the houses, fortifications, and inclosures, that remained. " The sea has, therefore, only resumed its former position at the base of the cliffs; the site of the old town having been an ancient bed of shingle, abandoned for ages by the ocean, perhaps contemporaneously with the retreat of its waters from the valley of the Ouse. Should the advancement of the sea be still progressive, Lewes Levels may again become an estuary, and the town of the *Cliff*, and the hamlet of *Landport*, regain the character from which their names were derived. See page 39.

 " Illustrations of the Geology of Sussex, page 292. Geology of the South-East of England, page 23. Dallaway's Western Sussex, Vol. I. page 55.

APPENDIX.

———

A. *Page* 17.—THE SURFACE OF THE MOON.—The moon is the only planetary body placed sufficiently near us, to have the inequalities of its surface rendered distinctly visible with the telescope. Attendant on the earth, and having nearly the same density, we may reasonably infer that the mineral substances of which it is composed do not differ essentially from those on the surface of our own planet. Astronomers now generally admit that the moon is surrounded by a very clear atmosphere, but which is so low that it scarcely occasions a sensible refraction of the rays of light when it passes over the fixed stars. Many of the dark parts of the moon, particularly the part called *Mare Crisium*, appear to be covered with a fluid, which may probably be more transparent and less dense than water, as the form of the rocks and craters are seen beneath it, but not so distinctly as in the lighter parts of the moon's surface. To examine the moon with a reference to its external structure, the defining power of the telescope should be of the first quality, sufficient to show the projections of the outer illuminated limb as distinctly as they appear when the moon is passing over the disk of the sun during a solar eclipse. With such a telescope, and a sufficient degree of light and of magnifying power, almost every part of the moon's surface appears to be volcanic, containing craters of enormous magnitude and vast depth: the shelving rocks, and the different internal ridges within them, mark the stations at which the lava has stood and formed a floor during different eruptions; while the cones in some of the craters resemble those formed within modern volcanoes. The largest mountain on the southern limb of the moon, like the largest volcanic cone on the earth, Chimborazo, has no deep crater on its summit. There are indeed the outlines of the crater, but it is nearly filled up; while from the foot of this lunar mountain diverging streams of lava seem to flow in different directions, to the distance of six hundred miles. The longest known current of modern lava on the earth is in Iceland; it extends sixty miles; but the volcanoes in that island bear no proportion to those of the moon in magnitude.—*Mr. Bakewell.*

B. *Page* 52.—THE LAKE OF THE SOLFATARA.—Its temperature was, in the winter, in the warmest parts, above 80 deg. of Fahrenheit, and it appears to be pretty constant; for I have found it differ a few degrees only, in January, March, May, and the beginning of June; it therefore being nearly twenty degrees above the mean temperature of the atmosphere, must be supplied with heat from a subterraneous source. Kircher has detailed in his *Mundus Subterraneus* various wonders respecting this lake, most of which are unfounded, such as that it is unfathomable, that it has at the bottom the heat of boiling water, and that floating islands rise from the gulf. It must certainly be very difficult, or even impossible to fathom a source which rises with so much violence from a subterraneous excavation; and at a time when chemistry had made small progress, it was easy to mistake the disengagement of carbonic acid for an actual ebullition. The floating islands are real, but neither the Jesuit nor any of the writers who have since described this lake, have had a correct idea of their origin, which is exceedingly curious. The high temperature of this water, and the quantity of carbonic acid that it contains, render it peculiarly fitted to afford a pabulum or nourishment to vegetable life; the banks of travertine are every where covered with reeds, lichens, confervæ, and various kinds of aquatic vegetables. At the same time that the process of vegetable life is going on, the crystallizations of the calcareous matter, which is everywhere deposited in consequence of the escape of carbonic acid, likewise proceeds, and gives a constant milkiness to what from its tint would otherwise be a blue fluid. So rapid is the vegetation, owing to the decomposition of the carbonic acid, that even in winter masses of confervæ and lichens, mixed with deposited travertine, are constantly detached by the currents of water from the bank, and float down the stream, which being a considerable river, is never without many of these small islands on its surface. They are sometimes only a few inches in size, and composed merely of dark green confervæ, or purple or yellow lichens; but, occasionally, are even several feet in diameter, and contain seeds and various species of common water-plants, which are usually more or less incrusted with marble. There is, I believe, no place in the world where there is a more striking example of the opposition or contrast of the laws of animate and inanimate nature, of the forces of inorganic chemical affinity, and those of the powers of life. Vegetables, in such a temperature, and everywhere surrounded by food, are produced with a wonderful rapidity; but the crystallizations are formed with equal quickness, and are no sooner produced than they are destroyed together. Notwithstanding the sulphureous exhalations from the lake, the quantity of vegetable matter generated there, and its heat, make it the resort of an infinite variety of insect tribes; and, even in the coldest days in winter, numbers of flies may be observed on the vegetables surrounding its banks, or on its floating islands. Their larvæ may also be seen there, sometimes incrusted and entirely destroyed by calcareous matter, as well as the insects themselves, and various species of shell-fish that are found amongst the vegetables which grow and are destroyed in the travertine on its banks. Snipes, ducks, and other water-birds, often visit these lakes, probably attracted by the temperature and the quantity of food in which they abound; but they usually confine themselves to the banks, as the carbonic acid disengaged from the surface would be fatal to them, if they ventured to swim upon it when tranquil. In May 18—, I fixed a stick on a mass of travertine covered by the water, and examined it in the beginning of the April following, for the purpose of determining the nature of the depositions. The water was lower at this time; yet I had some difficulty, by means of a sharp-pointed hammer, in breaking the mass which adhered to the bottom of the stick; it was several inches in thickness. The upper part was a mixture of light tufa and leaves of confervæ; below this was a darker and more compact travertine, containing black and decomposed masses of confervæ; in the inferior part, the travertine was more solid, and of a grey colour, but with cavities which I have no doubt were produced by the decomposition of vegetable matter. I have passed many hours, I may say days, in studying the phenomena of this wonderful lake; it has brought trains of thought into my mind connected with the early changes of our globe; and I have sometimes reasoned from the forms of plants and animals preserved in marble in this thermal source, to the grander depositions in the secondary rocks, where the zoophytes or coral insects have worked upon a grand scale, and where palms and vegetables, now unknown, are preserved with the remains of crocodiles, turtles, and gigantic extinct saurian animals, which appear to have belonged to a period when the whole globe possessed a much higher temperature. I have likewise often been led, from the remarkable phenomena surrounding me in that spot, to compare the works of man with those of nature. The baths, erected there nearly twenty centuries ago, present only heaps of ruins, and even the bricks of which they were built, though hardened by fire, are crumbled into dust; whilst the masses of travertine around, though formed by a variable source from the most perishable materials, have hardened by time, and the most perfect remains of the greatest ruins in the eternal city, such as the triumphal arches and the Colosseum, owe their duration to this source.

How marvellous are those laws by which the humblest types of organic existence are preserved, though born amidst the sources of their destruction, and by which a species of immortality is given to generations floating, as it were, like evanescent bubbles on a stream raised from the deepest caverns of the earth, and instantly losing what may be called its spirit in the atmosphere.—*Sir Humphrey Davy's Last Days of a Philosopher.*

———

C. *Page* 54.—CAVERNS.—One of the most common appearances in limestone caverns, is the formation of what are called *stalactites,* from a Greek word, signifying distillation, or dropping. To explain these, a brief description of the mode of their production will be necessary. Whenever water filters through a limestone rock, it dissolves a portion of it; and on reaching any opening, such as a cavern, either at its sides or roof, it forms a *drop*, the moisture of which is soon evaporated by the air, leaving a small circular *plate* of calcareous matter; another drop succeeds in the same place, and adds, from the same cause, a fresh coat of incrustation. In time, these successive additions produce a long, irregular, conical projection from the roof, which is continually being increased by the fresh accession of water loaded with calcareous or chalky matter, which it deposits on the outside of the *stalactite* already formed, and trickling down, adds to its length by subsiding to the point, and being dried up as before; precisely in the same manner as during frosty weather, icicles, which are *stalactites of ice*, or frozen water, are formed on the edge of the eaves of a roof. When the supply of water holding lime in solution, is too rapid to allow of its evaporation at the bottom of the *stalactite*, it drops to the floor of the cave, and drying up gradually, forms, in like manner, a *stalactite* rising upwards from the ground, instead of hanging from the roof; these are called, for the sake of distinction, *stalagmites*.

It frequently happens, where these processes are uninterrupted, that a *stalactite* hanging from the roof, and a *stalagmite* formed immediately under it, from the superabundant water, increase till they unite, and thus constitute a natural pillar, apparently supporting the roof of the grotto; it is to the grotesque forms assumed by stalactites, and these natural columns, that caverns owe the interesting appearances, described in such glowing colours by those who witness them for the first time.—*Saturday Magazine*, No. 42.

D. *Page* 54.—Weyer's Cave.—This cave is situated in a ridge of limestone hills, running parallel to the blue mountains. A narrow and rugged fissure leads to a large cavern, where the most grotesque figures, formed by the percolation of water through beds of limestone, present themselves, while the eye, glancing onward, watches the dim and distant glimmers of the lights of the guides—some in the recess below, and others in the galleries above. Passing from these recesses, the passage conducts to a flight of steps that leads into a large cavern of irregular form, and of great beauty. Its dimensions are about thirty feet by fifty. Here the incrustations hang just like a sheet of water that has been frozen as it fell; there they rise into a beautiful stalactitic pillar, and yonder compose an elevated seat, surrounded by sparry pinnacles. Beyond this room is another, more irregular, but more beautiful. Besides having sparry ornaments in common with the others, overhead is a roof of the most admirable and singular formation. It is entirely covered with *stalactites*, which are suspended from it like inverted pinnacles. They are of the finest material, and are most beautifully shaped and embossed. In another apartment, an immense sheet of transparent *stalactite* extends from the roof to the floor, which, when struck, emits deep and mellow sounds, like those of a muffled drum. Farther on is another vaulted chamber, which is one hundred feet long, thirty-six wide, and twenty-six high. Its walls are filled with grotesque concretions. The effect of the lights placed by the guides at various elevations, and leaving hidden more than they reveal, is extremely fine. At the extremity of another range of apartments, a magnificent hall, two hundred and fifty feet long, and thirty-three feet high, suddenly appears. Here is a splendid sheet of rock-work running up the centre of the room, and giving it the aspect of two separate and noble galleries; this partition rises twenty feet above the floor, and leaves the fine span of the arched roof untouched. There is a beautiful concretion here, which has the form and drapery of a gigantic statue; and the whole place is filled with stalagmitical masses of the most varied and grotesque character. The fine perspective of this room, four times the length of an ordinary church, and the amazing vaulted roof spreading overhead, without any support of pillar or column, produces a most striking effect. In another apartment, which has an elevation of fifty feet, there is at one end an elevated recess, ornamented with a group of pendant *stalactites* of unusual size, and singular beauty. They are as large as the pipes of a full-sized organ, and ranged with great regularity; when struck, they emit mellow sounds of various keys, not unlike the tones of musical glasses. Other cavities, profusely studded with sparry incrustations, extend through the limestone rock. The length of this extraordinary group of caverns is not less than one thousand six hundred feet.—*Abridged from "A Narrative of the Visit to the American Churches," by Drs. Reed and Matheson.*

———

E. *Page* 70.—RECENT FORMATION OF SANDSTONE.— *From the Transactions of the Royal Geological Society of Cornwall, by Dr. Paris.*—" A sandstone occurs in various parts of the northern coast of Cornwall, which affords a most instructive example of a recent formation; since we here actually detect Nature at work in converting calca-reous sand into stone. A very considerable portion of the northern coast of Cornwall is covered with a calcareous sand, consisting of minute particles of comminuted shells, which, in some places, has accumulated in quantities so great, as to have formed hills of from forty to sixty feet in elevation. In digging into these sand hills, or upon the occasional removal of some part of them by the winds, the remains of houses may be seen: and in some places, when the churchyards have been overwhelmed, a great number of human bones may be found. The sand is supposed to have been originally brought from the sea by hurricanes, probably at a remote period. At the present moment, the progress of its incursion is arrested by the growth of the *arundo arenacea.* The sand first appears in a slight but increasing state of aggregation on several parts of the shore in the Bay of St. Ives; but, on approaching the Gwythian river, it becomes more extensive and indurated. On the shore opposite Godrevy Island, an immense mass of it occurs, of more than a hundred feet in length, and from ten to twenty in depth, containing entire shells and frag-ments of clay-slate; it is singular that the whole mass assumes a striking appearance of stratification. In some places, it appears that attempts have been made to separate it, probably for the purpose of building, for several old houses in Gwythian are built of it. The rocks in the vicinity of this recent formation in the Bay of St. Ives, are greenstone and clay slate, alternating with each other. The clay slate is in a state of rapid decomposition, in consequence of which large masses of the Hornblende rock have fallen in various directions, and given a singular character of picturesque rudeness to the scene. This is remarkable in the rocks which constitute Godrevy Island. It is around the promontory of New Kaye, that the most extensive formation of sandstone takes place. Here it may be seen in different stages of induration, from a state in which it is too friable to be detached from the rock upon which it reposes, to a hardness so considerable that it requires a very violent blow from a sledge to break it. Buildings are here constructed of it; the church of Cranstock is entirely built with it; and it is also employed for various articles of domestic and agricultural uses. The geologist who has previously examined the celebrated specimen from Guadaloupe, will be struck with the great analogy which this formation bears to it. Suspecting that masses might be found containing human bones imbedded, if a diligent search were made in the vicinity of those cemeteries which have been overwhelmed, I made some investigations in those spots, but, I regret to add, without success. The rocks upon which the sandstone reposes, are alternations of clay slate, and slaty limestone. The inclination of the beds is SS.W., and at an angle of 40º. Upon a plane formed by the edges of these strata, lies a horizontal bed of rounded pebbles, cemented together by the sandstone which is deposited immediately above them, forming a bed of from ten to twelve feet in thickness, and containing fragments of slate, and entire shells; and exhibiting the same appearance of stratification as that noticed in St. Ives Bay. Above this sandstone lie immense heaps of drifted sand. But it is on the western side of the promontory of New Kaye, in Fishel Bay, that the geologist will be most struck with the formation; for here no other rock is in sight. The cliffs, which are high, and extend for several miles, are entirely composed of it; they are occasionally intersected by veins and dykes of breccia. In the cavities, calcareous stalactites of rude appearance, opaque, and of a grey colour, hang suspended. The beach is covered with disjointed fragments, which have been

detached from the cliffs above, many of which weigh two
or three tons."

———

F. *Page* 83.—Lithodomi, or Boring Mollusca; *which
have the power of perforating rocks.*—Every one who has
walked by the sea-side must have observed the blocks
and masses of the chalk rocks full of perforations; and if
his curiosity have induced him to examine these with attention, he will have perceived that though many of the cavities
are empty, some of them contain the shelly remains of the
animals which once inhabited them. The power possessed
by creatures so delicate, and with such fragile coverings,
of excavating the solid rock, has naturally excited much
speculation as to the mode by which the perforations are
effected; and it is now generally admitted, that it is not
by mechanical power only that the feeble inhabitants of
the boring shells are able to form themselves a secure
asylum in the rock, but by the secretion of a liquid which
acts chemically on the stone, softens it, and renders it
capable of being removed with facility. In a late volume
of the Philosophical Transactions, there is an interesting
paper on the economy of molluscous animals, by Mr. Gray,
which throws much light on this subject. It appears that,
although *teredines, pholades,* and other boring shells, are
covered with short spines and striæ, by means of which
they were supposed capable of rasping stones, yet other
mollusca which inhabit stony cavities are perfectly smooth.
On the shore, near Kemptown, a *pholas,* which has a
rasping apparatus, and a *venus,* wholly destitute of a rugous
surface, may be seen in cavities of the chalk. Shells
of this kind have not been observed to bore into any other
substances (wood excepted) than shells, marl, chalk, limestone, and sandstone, consolidated by calcareous cement.
Granite appears to resist all the dissolving powers of the
mollusca. Thus, in the Plymouth Breakwater, in which
limestone and granite are employed and placed side by
side, the *patellæ,* or limpets, form their rounded holes in
the former, while they do not in the slightest degree alter
the surface of the latter, except by clearing off from it
any adherent calcareous substance.

G. *Page 85.*—OBSERVATIONS ON THE TEMPLE OF SERA-
PIS AT PUZZUOLI, NEAR NAPLES; *in a Letter to W. H. Fitton,
M.D., from Charles Babbage, Esq.*—This paper commences with a general description of the present state of
the Temple of Serapis, and gives the measurement of the
three marble columns which remain standing, and which,
from the height of eleven feet to that of nineteen, are perforated on all sides by the *Modiola lithophaga* (of Lamarck);
the shells of that animal remaining in the holes formed by
them in the columns. A description follows of the present
state of twenty-seven portions of columns, and other fragments of marble, and also of the several incrustations
formed on the walls and columns of the temple.

From these and other data, Mr. Babbage concludes :—

1. That the temple was originally built at, or nearly at
the level of the sea, for the convenience of sea-baths, as
well as for the use of the hot spring which still exists on
the land side of the temple.

2. That, at a subsequent period, the ground on which
the temple stood, subsided slowly and gradually ; the salt
water, entering through a channel which connected the
temple with the sea, or by infiltration through the sand,
mixed itself with the water of the hot spring containing
carbonate of lime, and formed a lake of brackish water in
the area of the temple, which, as the land subsided, became
deeper, and formed a dark incrustation.

The proofs are, that sea-water alone does not produce a
similar incrustation ; and that the water of the hot spring
alone produces an incrustation of a different kind ; also,
that Serpulæ are found adhering to this dark incrustation;
and that there are lines of water-level at various heights
from 2.9 feet to 4.6 feet.

3. The area of the temple was now filled up to the height
of about seven feet with ashes, tufa, or sand, which stopped
up the channel by which sea-water had been admitted.
The waters of the hot spring thus confined, converted the
area of the temple into a lake, from which an incrustation
of carbonate of lime was deposited on the columns and walls.
The proofs are, that the lower boundary of this incrustation
is irregular ; whilst the upper is a line of water-level, and
that there are many such lines at different heights;—that
salt water has not been found to produce a similar incrus-

tation;—that the water of the Piscina Mirabile, which is distant from the sea, but in this immediate neighbourhood, produces, according to an examination by Dr. Faraday, a deposit almost precisely similar;—that no remains of Serpulæ, or other marine animals, are found adhering to it.

4. The temple continuing to subside, its area was again partially filled with solid materials; and at this period was subjected to a violent incursion of the sea. The hot-wa'er lake was filled up, and a new bottom produced, entirely covering the former, and concealing also the incrustation of carbonate of lime.

The proofs are, that the remaining walls of the temple, are highest on the inland side, and decrease in height towards the sea-side, where they are lowest;—that the lower boundary of the space perforated by the marine *Lithophagi* is, on different columns, at different distances beneath the uppermost, or water-level line;—that several fragments of columns are perforated at the ends.

5. The land continuing to subside, the accumulations at the bottom of the temple were submerged, and modiolæ attaching themselves to the columns and fragments of marble, pierced them in all directions. The subsidence continued until the pavement of the temple was at least nineteen feet below the level of the sea.

The proofs are derived from the condition of the columns and fragments.

6. The ground on which the temple stood, appears now to have been stationary for some time, but it then began to rise. A fresh deposition of tufa, or of sand, was lodged, for the third time, within its area, leaving only the upper part of three large columns visible above it.

Whether this took place before or subsequently to the rise of the temple to its present level, does not appear; but the pavement of the area is at present level with the waters of the Mediterranean.

The author then states several facts, which prove that considerable alterations in the relative level of the land and sea have taken place in the immediate vicinity. An ancient sea-beach exists near Monte Nuovo, two feet above the present beach of the Mediterranean. The broken columns of the Temples of the Nymphs and of Neptune, remain at present standing in the sea. A line of perforations of modiolæ, and other indications of a water-level four feet above the present sea, are observable on the sixth pier of the bridge of Caligula; and again on the twelfth pier, at the height of ten feet. A line of perforations by modiolæ is visible in a cliff opposite the island of Nisida, thirty-two feet above the present level of the Mediterranean. —*Abstract of the Proceedings of the Geological Society; March, 1834.*

H. *Page* 110.—Irish Elk; *Cervus Megaceros.*—Beds of gravel and sand, containing recent marine shells and bones of the Irish Elk, have been observed by Dr. Scouler in the vicinity of Dublin, at an elevation of two hundred feet above the level of the sea. It is therefore manifest that this extinct quadruped, although found in peat-bogs and morasses at a comparatively recent period, must have been an inhabitant of Ireland antecedently to some of the last changes in the relative position of the land and water. The discovery of a vast number of skeletons of the Elk in the small area of the Isle of Man, seems to indicate a great alteration in the extent of land and sea; for it is difficult to conceive that such herds of this gigantic race could exist in so limited a district; and it is therefore probable that the island was separated from the main land at no remote geological period by subsidences commensurate with the elevation of which Ireland affords such decisive evidence. —*Address of Charles Lyell, Esq., President of the Geological Society of London, 1837.*

I. *Page* 312.—A Tabular Arrangement of the Fossil Fishes of the Chalk Formation of the South-East of England, collected by Gideon Mantell, ll.d. f.r.s.

(From Recherches sur les Poissons Fossiles, by M. Agassiz.)

" Tout le monde sait que le Musée de M. le Dr. Mantell à Brighton est une collection classique pour la craie, et la formation Veldienne. Les soins minutieux que M. Mantell a donnés depuis bien des années à ces fossiles, les ont rendus plus parfaits que tous ceux des autres musées : car souvent il est parvenu à les detacher entièrement de la roche dans laquelle ils se trouvaient, ou du moins à les produire en relief, en détachant toutes les matières solides qui recouvraient les parties les mieux conservées de l'animal."

ORDER I.—The *Placoidians*, (from πλαξ, a broad plate.) The skin, covered irregularly with enamelled plates, sometimes of a large size, but frequently in the form of small points, as in the shagreen on the skin of Sharks, and the tubercles on the integuments of Rays.

PTYCHODUS *latissimus.* Mantell's South Down Fossils. Tab. xxxii. fig. 19.
        Agassiz, Poiss. Foss. Vol. iii. tab. 25.
——————*polygyrus.* Ibid. Tab. xxxii. fig. 23, 24.
——————*mammillaris.* Ibid. Tab. xxxii. fig. 18, 20, 25, 29.
——————*decurrens.*
——————*altior.* South Down Fossils. Tab. xxxii. fig. 17, 21, 27.

Teeth, and perhaps vertebræ, of the above species, and a few examples of their dorsal defences, (*Ichthyodorulites* of Dr. Buckland,) are the only remains hitherto discovered. (*Agass. Poiss. Foss.* Vol. iii. tab. 10ª, 10ᵇ.) The teeth were referred to fishes of the genus Diodon, by previous authors, and the defences were called radii, or fin-bones of Balistes, and Siluri.

Teeth of a new species of *Ptychodus* have been discovered in the sand of New Jersey, United States, by Dr. Morton.—(*Morton's Synopsis*, Pl. 18, fig. 1, 2.) I have named it *Ptychodus Mortoni.*

PTYCHODUS,—*spec. undetermined.* Dorsal defences, and a beautiful example of a fin, are represented in the fossils of the South
        Downs. Tab. xxxiv. fig. 8. Tab. xxxix. and Tab. xl. fig. 3.
GALEUS *pristodontus.* South Down Fossils. Tab. xxxii. fig. 12 to 16.
        Agass. Poiss. Foss. Vol. iii. tab. 26. fig. 14.
NOTIDANUS *microdon.* Ibid. Tab. xxxii. fig. 22.
LAMNA *appendiculata.* Ibid. Tab. xxxii. fig. 2, 3, 5, 6, 9.
——————*acuminata.* Ibid. Tab. xxxii. fig. 1.
——————*Mantellii.* Ibid. Tab. xxxii. fig. 4, 7, 8, 10.
——————*crassissima.* Not figured.
ODONTAPSIS *raphiodon.* Not figured.
SPINAX *major.* Agass. Poiss. Foss. Vol. iii. tab. 10. fig. 8, 14.
PSAMMODUS *asper.* Poiss. Foss. Vol. iii. tab. 10. fig. 1, 3.
ACRODUS *transversus.* Poiss. Foss. Vol. iii. tab. 10. fig 4, 5.
GYRODUS *angustus.* Poiss. Foss. Vol. ii. tab. 66ª. fig. 14, 15.

The above order of fishes is represented by five genera, of which one, containing twelve species, is extinct. The fishes of the genera Ptychodus, Galeus, and Lamna, are very widely distributed.

ORDER II.—The *Ganoidians,*—(γανος, splendour, from the brilliant surface of their enamel.) These are characterised by angular scales, formed of horny or bony plates, protected by a thick layer of enamel.

MACROPOMA *Mantellii.* Tab. 38, page 308. Length 24 inches. South
Down Fossils. Tab. xxxvii. and xxxviii.
Agass. Poiss. Foss. Vol. v. tab. 60b. fig. 2.
———————————— Coprolites of: South Down Fossils. Tab. ix.
fig. 5, 11. Agass. Poiss. Foss. Vol. ii. tab. 65.

This Macropoma is perhaps the most remarkable of all
the fossil fishes; in most examples the membranes of the
stomach are preserved.

SPHŒRODUS *mammillaris.* Not figured. From Clayton chalk-pit.
DERCETIS *elongatus.* Tab. 39, page 309. Length 16 inches. South Down
Fossils. Tab. xxxiv. fig. 10, 11. Tab. xl. fig. 2.
Agass. Poiss. Foss. Vol. ii. tab. 66ᵃ. fig. 1 to 8.

The above order comprehends three extinct genera, with
three species. Another species of *Dercetis* has been found
in the chalk of Westphalia.

ORDER III.—The *Ctenoïdians,* (κτεις, a comb.) The
scales of this order are pectinated on their posterior margin,
like the teeth of a comb, and are composed of laminæ of
horn or bone, but have no enamel.

BERYX *Lewesiensis.* (B. ornatus of Agassiz.) Tab. xli. page 311. Length
12 inches. South Down Fossils. Tab. xxxiv. fig. 6.
Tab. xxxv. Tab. xxxvi. Agass. Poiss. Foss. Vol. iv.
tab. 14ᵃ.
——— *radians.* Tab. 40, page 310. Length 7 inches. History of the
County of Sussex. Vol. ii. Part ii. p. 15. fig. 22. Agass.
Poiss. Foss. Vol. iv. tab. 14b. fig. 7.
——— *microcephalus.* Agass. Poiss. Foss. Vol. iv. tab. 4c. fig. 7 to 9.

There are other species of Beryx in the chalk of Bohemia
and Westphalia; and genera nearly related to *Beryx,* in
the schist of Glaris. In England this order contains but
three species of a genus, of which we know but one living
species.

ORDER IV.—The *Cycloïdians,* (κυκλος, a circle.) The
scales smooth, with a simple margin, composed of laminæ
of horn or bone without enamel.

OSMEROIDES *Mantellii.* (*Salmo?* *Lewesiensis* of Mantell.) Tab. 37,
page 207. Length 12 inches. South Down Fossils.
Tab. xxx. fig. 12. Tab. xxxiv. fig. 3. Tab. xl. fig. 1.
Agass. Poiss. Foss. Vol. v. tab. 60º.

To the above species belong the remarkable uncompressed specimens in my Museum.

OSMEROIDES *Lewesiensis.* Agass. Poiss. Foss. Vol. v. tab. 60c. (*Salmo?*
        Lewesiensis of Mantell.) This species is more elon-
        gated than *O. Mantellii,* and the number of rays in
        the dorsal fin is greater.
————— *granulatus.* History of Lewes. Vol. i. plate xxix. fig. 13.
        The bones of the head, with the jaws and teeth, have
        alone been discovered. Agass. Poiss. Foss.
ENCHODUS *halocyon.* South Down Fossils. Tab. xxxiii. fig. 2, 3, 4.
        Tab. xliv. fig. 1, 2. Agass. Poiss. Foss. Vol. v. tab. 25.
        fig. 11 to 6.
SAUROCEPHALUS *lanciformis.* (Harlan.) South Down Fossils. Tab.
        xxxiii. fig. 7, 6. Trans. Geol. Soc. of Pennsylvania,
        vol. i. p. 83. Agass. Poiss. Foss. Vol. v. tab. 25. fig.
        21 to 29.
SAURODON *Leanus.* (Hays.) Trans. American Philos. Society, vol. for
        1830, plate 16. Agass. Poiss. Foss. Vol. v. tab. 25. fig.
        17 to 20.
HYPSODON *Lewesiensis.* South Down Fossils. Tab. xxxiii. fig. 8. Tab.
        xlii. fig. 1 to 5. Agass. Poiss. Foss. Vol. v. tab. 25a.

From the resemblance of the teeth of this fish, to those
of reptiles, it was supposed that the original belonged to
an extinct genus of saurians; but in 1833, a considerable
portion of the head, with the maxillæ, many vertebræ, &c.,
were discovered in a block of chalk, near Lewes, and the
true characters of this remarkable ichthyolite determined.

    ⁎ The following fishes have been named by M. Agassiz,
since the above table was constructed.

ACROGNATHUS *boops.* Tab. 37, page 307. Natural size. Agass. Poiss.
        Foss. Vol. iii. tab. 60a. fig. 1, 4. An unique specimen,
        from Southerham quarry, near Lewes.
AULOLEPIS *typus.* Tab. 38, page 308. Length 6 inches. An unique
        specimen, from Clayton chalk-pit, Sussex. One
        nearly perfect example has alone been found. Poiss.
        Foss. Vol. iii. tab. 60. fig. 5, 8.
BELONOSTOMUS *cinctus.* Agass. Poiss. Foss. Vol. ii. tab. 66a. fig. 10 to 13.
CHIMERA *Agassizii.* Agass. Poiss. Foss. Vol. iii. pl. 40. fig. 3, 5. (De-
        termined by Dr. Buckland.) The beaks or mandibles
        have alone been discovered.
————— *Mantellii.* Agass. Poiss. Foss. Vol. iii. pl. 40. fig. 1, 2. Two
        mandibles were found, many years since, in a block
        of chalk, near Lewes. This species also occurs in the
        Shanklin sand of Kent. A beak has been found by
        Mr. W. H. Bensted in the Iguanodon quarry, near
        Maidstone.
TETRAPTERUS *minor.* Lewes. Agass. Poiss. Foss. Vol. iii. tab. 60, fig.
        1, 4.
CATURUS *similis.* Agass. Poiss. Foss. Vol. ii. tab. 66a. fig. 9.
ACROTEMNUS *faba.* Poiss. Foss. Vol. ii. tab. 66a. fig. 16, 18.

GLOSSARY.

Explanations of many scientific terms, not inserted in the Glossary, will be found in the text, by consulting the Index.

Acephala............ Molluscous animals without a head, as the oyster, &c.

Algæ .... A family of sea-weeds.

Alluvium .... Waterworn materials.

Aluminum .... Metallic base of clay.

Amorphous .... Shapeless.

Amygdaloid........ Cellular volcanic rock, the cavities of which are filled with other substances.

Antennæ .... The feelers of insects.

Anthracite .... Stone or cannel coal.

Anthracotherium . An extinct animal, allied to the Palæotheria, found in Anthracite.

Arenaceous .... Formed of sand.

Argillaceous .... Formed of clay.

Astrea .... A genus of corals.

Augite .... A mineral found in many volcanic rocks.

Basalt .... Ancient lava, composed of Augite and Felspar, frequently columnar.

Basin .... A depression of, or concavity in, strata.

Belemnite............ The bone of an animal allied to the cuttle-fish.

Bitumen .... Mineral pitch or tar.

Blende .... Sulphuret of zinc, occurring in primary and secondary rocks.

Branchiæ............ The respiratory apparatus of aquatic animals.

Breccia............... Conglomerate of pebbles or fragmented rocks.

Calc sinter .... Deposition from thermal springs charged with carbonate of lime.

Calcaire Grossier. A tertiary limestone.

Calcium .... Metallic base of lime.

Campanulariæ. ... Arborescent corals, with bell-shaped cells ; p. 475.

Carbon .... The elementary substance of charcoal and the diamond.

Carbonate of Lime. Lime and carbonic acid.

Carboniferous...... Belonging to coal.

Cephalopoda .... Mollusca, with the organs of motion around their heads.

Caryophillia .... Branched stellular coral.

Centrifugal .... A force directed from the centre to the circumference.

Cetacea............... Marine mammalia, as the whale, porpoise, &c.
Chalcedony .... A species of silex, named from Chalcedon, a city of
        Asia, near which it is found in great abundance.
Chert.................... A silicious mineral allied to flint and chalcedony.
Choanite .... A zoophyte of the chalk.
Cilia .... Hair-like vibratory organs.
Cirrus .... A fossil shell of the chalk. See Geol. S. E. E., p. 125.
Cornbrash............. A coarse shelly limestone.
Conchoidal .... Shelly.
Concretion .... A coalition of separate particles.
Coleoptera .... Insects having wing-cases, as beetles.
Conformable .... Applied to parallel strata lying upon each other.
Condyle .... An articulating surface or joint.
Conglomerate ...... Fragments cemented together.
Coniferæ .... Trees bearing cones, as the fir, pine, &c.
Cotyledons .... Seed-lobes of plants.
Crag .... A tertiary deposit; from a provincial term (used in
        Suffolk and Norfolk) to denote gravel.
Crater .... The vent of a volcano.
Crateriform......... Having the form of a crater.
Crinoidea............ Lily-shaped animals.
Crustacea............ Animals allied to the crab, lobster, &c.
Cryptogamia .... Plants with concealed fructification, as mosses, ferns,
        &c.
Crystalline .... Presenting the structure of crystals.
Crystals .... Symmetrical forms assumed by mineral substances.
Cyathiform .... Cup-shaped.
Cycadea .... A genus of plants allied to the palms and ferns.

Delta .... Alluvial deposites formed by rivers.
Denudation......... Strata exposed by the action of water.
Detritus .... Disintegrated materials of rocks.
Dicotyledonous ... Plants with seeds having two lobes.
Didelphis........... A marsupial animal, allied to the Opossum.
Diluvium........... A term formerly employed to designate ancient
        alluvial deposites.
Dip .... The inclination of strata.
Diptera .... Insects having two wings.
Discoidal........... In the form of a disk.
Dyke................. An intrusion of melted matter into rents or fissures of
        solid rocks.

Earth's Crust ...... That portion of the solid surface of the earth which
        is accessible to human observation.
Echinus .... Sea-urchin.
Elytra .... Wing-cases of insects.
Encrinite............ A genus of crinoidea.
Eocene .... The dawn of the present epoch; the early tertiary.
Ephemeron .... The creature of a day.
Escarpment........ The steep cliff of a ridge of land.
Exuviæ .... Organic remains.

Fault................. Interruption of the continuity of strata with displace-
        ment.
Fauna .... The zoology of a particular country.
Felspar.............. A mineral which enters into the composition of many
        primary rocks.
Ferruginous ...... Impregnated with iron.
Flora .... The botany of a particular country.
Flustra.............. A genus of Polyparia; p. 460.

GLOSSARY.

Formation .... Formation, group, series, system; terms applied *to* a class of rocks, presumed to have been formed during one geological epoch.

Fungia............... A genus of corals; p. 482.

Galt .... A provincial term applied to the blue marl of the chalk formation.

Gelatinous .... Like jelly.

Gorgonia .... A flexible arborescent species of Polyparia; p. 475.

Gneiss .... A primary rock, allied to granite.

Granite............... A primary rock, composed of mica, quartz, and felspar.

Greensand, or⎱ The lowermost strata of the chalk formation.
Shanklin Sand⎰

Greenstone .... A trap rock.

Greywacke .... A transition rock ; a conglomerate indurated by heat.

Grit .... Coarse-grained calciferous sahdstone.

Gypsum .... Sulphate of lime.

Hamites .... Hook-shaped shells, of a genus of Cephalopoda.

Hemiptera .... Insects with wings, half horny and half membraneous.

Hylæosaurus ...... The Wealden lizard.

Hymenoptera...... Insects with membraneous wings.

Icthyosaurus ...... Fish-like lizard.

Iguanodon .... Extinct colossal lizard, having teeth like those of the Iguana.

Imbricated .... Laid one over the other like scales.

Induction .... The derivation of a principle from facts.

Infusoria............ Microscopic animals, found in infusions, &c.

Inoceramus .... A bivalve shell of the chalk, having a fibrous structure. See Geol. S. E E., p. 127.

Invertebrated Animals. Animals without vertebræ, as worms, &c.

Kimmeridge Clay. A blue clay of the Oolite.

Lacustrine .... Belonging to a lake.

Lamellated .... Covered with thin plates or scales.

Laminæ .... The thin layers of which a stratum is composed.

Laminated .... Strata formed of very thin layers.

Lapilli .... Volcanic ashes, in which minute globular concretions prevail.

Lepidoptera......... Insects with scaly wings.

Lias .... A provincial term, applied to a group of strata situated between the Oolite and the New Red Sandstone.

Lignite............... Carbonized wood.

Lithodomi .... Mollusca which perforate stone.

Lithological ...... The stony character of a mineral mass.

Lithophytes......... Stony plants ; a term applied to corals.

Littoral .... Belonging to the shore.

Loess.................. An alluvial tertiary deposit on the banks of the Rhine.

Lophiodon .... Fossil animal, allied to the Tapir, so named from eminences on the teeth.

Lycopodiacea ...... The family of Club Mosses.

Madrepore .... A genus of stellular corals,

Mammalia .... Animals which give suck to their young.

Mammillated ...... Studded with mammillæ, or rounded protuberances.

Mammoth .... An extinct genus of quadruped, allied to the elephant.

Marl.................. A mixture of clay and lime.

Marsupial .... Applied to animals having a pouch, as the kangaroo.

Marsupite .... A genus of crinoidea of the chalk.

GLOSSARY.

Mastodon .... An extinct genus of quadrupeds, allied to the elephant, having tuberculated teeth.
Matrix .... The substance in which a fossil is imbedded.
Meandrina .... A genus of corals, with meandering cells, as the brain-stone coral.
Megalosaurus ...... Gigantic lizard ; an extinct Saurian, allied to the Monitor.
Megalonyx .... An extinct quadruped, allied to the sloth.
Meteorites .... Mineral masses which fall from the atmosphere.
Megatherium ...... An extinct gigantic quadruped, allied to the sloth.
Mica .... A simple mineral, one of the component parts of granite.
Miocene .... Middle Tertiary series.
Molares .... Grinding teeth.
Mollusca .... Soft animals—destitute of a bony structure.
Monitor .... A genus of lizards, inhabiting the tropics.
Monocotyledonous Plants having seeds with but one lobe.
Multilocular ...... Many-chambered shells, as the Nautilus.
Muschelkalk .... A limestone of the red sandstone formation.

New Red Sandstone A group of strata between the magnesian limestone and the lias.
Nodule .... A rounded mineral mass, as a flint.
Normal .... The natural, or original condition.
Nucleus .... A centre, or point, round which other materials collect.
Nummulite .... An internal multilocular shell.

Obsidian .... A glassy lava.
Old Red Sandstone A group of strata lying between the carboniferous beds and the Silurian system.
Oolite .... A limestone group, composed of an aggregation of spheroidal grains.
Organic Remains . The relics of animals and plants.
Orthoceratite .... A straight, fossil multilocular shell
Ossicula .... Small bones.
Ovate .... Egg shaped.
Oxide .... The combination of oxygen with a metallic substance.

Pachydermata ...... Thick-skinned animals, as the rhinoceros, &c.
Palæontology ...... The science which treats of the structure of extinct animals.
Palæotherium .... An extinct quadruped, allied to the Tapir.
Paludina .... A genus of fresh-water mollusca.
Pectunculus .... A genus of bivalve shells.
Peperino .... A volcanic aggregate.
Petroleum .... Mineral oil.
Pisolite .... Minerals having a structure resembling peas agglutinated together.
Planorbis .... A genus of discoidal fresh-water shells.
Plastic clay .... A tertiary deposit.
Plesiosaurus .... An extinct marine animal, allied to the lizard.
Pliocene .... More recent, modern tertiary.
Polyparia .... Corals.
Porphyry .... A species of ancient lava.
Precipitate .... The chemical separation and deposit in a solid form of a substance held in solution by water.
Producta .... A genus of fossil, bivalve, marine shells, found in the lower secondary rocks, and allied to the Terebratula.
Pterodactyles ...... A genus of extinct winged reptiles.
Pumice .... A light spongy lava.
Pyriform .... Pear-shaped.
Pyrites .... Sulphuret of iron.

GLOSSARY.

Quartz .... A mineral composed of pure silex.

Ramose............... Branched.
Reticulated .... Resembling net-work.
Ruminants .... Animals which ruminate, as the ox, &c.

Saurian .... A division of the family of the lizards.
Scaphite .... An extinct genus of Cephalopoda, of a boat-like form.
Scoriæ .... Volcanic cinders.
Sedimentary .... Deposited by water.
Septaria .... Nodular masses of clay, having crevices filled with spar.
Sertularia .... A genus of arborescent corals.
Shale................... Slaty clay.
Silex .... Flint.
Silica .... The base of flint.
Silicious .... Flinty.
Silicified .... Changed to flint.
Silt .... Fluviatile deposites of mud, &c.
Spatangus .... A genus of the sea-urchin.
Spheroidal .... Having the form of a spheriod ; oblong, or oblate.
Spirifera .... An extinct genus of bivalve shells.
Stalactite............ Pendant masses of carbonate of lime.
Stalagmite .... Calcareous concretions formed on the floors of caverns by the dripping from stalactites.
Stellular .... Having star-like markings.
Stratum .... A layer of any deposit.
Stratified .... Deposited in layers.
Syenite................ A species of granite in which hornblende supplies the place of mica.

Tentacula .... Feelers.
Tertiary .... Applied to formations or strata newer than the chalk.
Testacea .... Shells.
Trachyte .... Lava composed chiefly of felspar.
Trap rocks .... Ancient volcanic rocks, the name of which is derived from a Swedish word, *Trappa*—a stair.
Trilobite .... An extinct family of Crustacea, the body being divided into three lobes, whence the name.
Tubipora .... Organ-pipe coral.
Tuff .... Earthy volcanic rock.
Tufa.................... A calcareous precipitate from water.
Turbinated .... Applied to shells having a spiral screw-shaped form.
Turrilite .... A spiral multilocular shell of the chalk ; p. 296.

Unconformable.... Applied to strata lying in a different plane to those on which they rest.

Veins.................. Fissures in rocks, filled up by mineral substances.
Vermes .... Worms.
Vertebrated......... Having a flexible osseous spinal column.
Vesicular............ Full of vesicles or cells.

Zoophytes .... Animal plants applied to corals, &c.

R. CLAY, PRINTER, BREAD-STREET-HILL.

Printed in the United States
By Bookmasters